Howard W. Sams and Company

INTERNET
GUIDE
TO THE
ELECTRONICS
INDUSTRY

Written and Compiled by
JOHN ADAMS

Howard W. Sams and Company

INTERNET
GUIDE
TO THE
ELECTRONICS
INDUSTRY

Written and Compiled by
JOHN ADAMS

A Division of Howard W. Sams & Company
A Bell Atlantic Company
Indianapolis, IN
http://www.hwsams.com

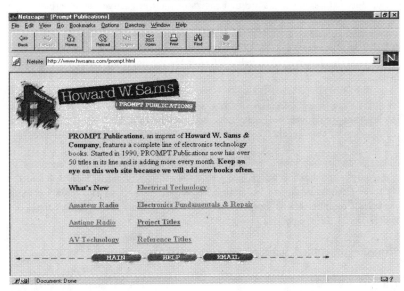

International Standard Book Number: 0-7906-1092-2

LIBRARY OF CONGRESS CATALOG CARD NO.: 96-72183

Acquisitions Editor: Candace M. Hall
Editor: Natalie F. Harris
Assistant Editors: Pat Brady, Loretta Leisure, Karen Mittelstadt
Typesetting: Natalie Harris
Pasteup: Suzanne Lincoln, Phil Velikan
Cover Design: Phil Velikan
Illustrations: Courtesy of Alpha-X Development, Alta Vista, EG3 Communications, Sam Engstrom, Excite Inc., Filip Gieszczykiewicz, Howard W. Sams & Co., and the Author.

Printed in the United States of America

9 8 7 6 5 4 3 2 1

CONTENTS

What is the Internet? — How Can the Internet Help an Electronics Hobbyist/Professional? — How Many People are On the Net? — Why are So Many People Talking About The Net? — What Is Available To View On the Internet? — What You Need To Get Online — How Much Is It? — How To Access the Net to Find Electronics Data If You Don't Have a Computer — How to Find What You Are Looking For — Is It Difficult To Learn the Internet? — What are the Terms of the Internet? — Do You Really Need the Internet?

What are the Electronics-Related Internet Resources? — Email — What Can You Send With Email? — What Does the "@" Symbol Mean, and What Is Your Address? — How You Can Use Email — Where To Get Programs For Email — How To Find Someone's Email Address — How To Find Out More About Email — FAQs - Frequently Asked Questions — What Questions Do Faqs Ask? — File Transfer Protocol — How To Find More Information About FTP — Internet Relay Chat — IRC Terms — What Channels Are Out There For Electronics and Related Topics? — Where to Find More Information and Programs for IRC — PDF - Portable Document Format — Additional Internet Resources

What Are Newsgroups? — What's On the UseNet for Electronics? — Example Text — Example Illustration — Example Direction to URL — What You Need to View Newsgroups — How to Use a Newsreader — Is There a Way to do Online Searches for Newsgroups or Articles? — Does it Cost Anything to Subscribe to Newsgroups? — How to Approach Newsgroups — What are Flaming and Spamming? — Hints to Keep the UseNet Free From Garbage — What do Alt, Comp and Sci Mean? — How to Use Newsgroups to Help Your Hobby/Profession

DEDICATION

To my father, John A. Adams, for instilling in me
his love of electronics.

To my brothers, Jeff, Jim, and Jason,
for our shared love of computers.

To my mother, Pam, and to my entire family
for their support.

To Kristy, for her love.

In memory of John A. Adams, Jeff Adams,
and Scott Bailey.

CREDITS

Thanks to Kristy Klein, Pam Cassa, Jim Adams, Jason Adams, Jennifer and Jordan Adams, Sharon and Darryl Baillargeon, and Mike Bailey.

To Mike Jackson at Aardvark Design.
— http://aardvark-design.com/

To the Collective Mind software company.
— http://theCollectiveMind.com/ or http://pobox.com/~tcm/

To Netlaw.
— http://pobox.com/~netlaw/ or http://www.netlaw.ca/

To InfiNet-FX — http://pobox.com/~ifx/

To Don Lancaster — http://www.tinaja.com/

To Devora Albelda, and Mark Plancke.

To Sam J. Engstrom at SEMS electronics.
— http://www.elixir.org/users/seng and http://www.elixir.org/SEMS_electronics

To Philip Lalone at Alpha-X Development.
— http://www.alphax.com/

To Werner Terreblanche, and to all others who responded to postings and Emails, or who otherwise helped in the compilation of this book and directory.

Thanks to Candace Drake Hall, Natalie Harris, and the rest of the Prompt Publications staff.

My personal Email is electronics@pobox.com or john.adams@pobox.com, and my personal URL is http://pobox.com/~electronics. Please feel free to send me any comments or suggestions regarding this book or electronics in general.

ABOUT THE AUTHOR

John J. Adams is co-owner of InfiNet-FX, an Internet consulting service for businesses, specializing in the electronics industry on the Web. Computers, electronics, writing, and education have always been among John's driving forces in life. It was only natural for him to create a book such as this, to help industry members and hobbyists who want to find out more about the vast, new information age. John has studied computers and electronics for 20 years, ever since his father got him interested at the age of 8. Now he teaches classes on Basic Electronics, as well as Computing and Internet Basics, and provides companies with alternative advertising methods. John Adams can be reached any time at *electronics@pobox.com* for questions or comments, or to help you get your company started and thriving on the Internet.

PREFACE

INTERNET mania has arrived. "Visit us Online at www. company.com," or "Download this Freeware Program on the Internet," is posted over our TV screens, magazines and newspapers nightly. Electronics companies, chip manufacturers, hobbyists, technicians, and electrical engineers are coming to the same conclusion — the Net is the ultimate research, entertainment and advertising tool. Where are we all headed? Into Cyberspace!

The Internet has evolved tremendously throughout its history. What started off as a government plan to communicate in times of war has turned into a near-monopolization of transglobal communications. What began as a research tool for universities and government agencies has become an instrument that gives the entire population of the earth access to the sum of human knowledge.

Now that PCs are in millions of homes and contain a reliability previously unknown, the natural progression is to attach them all together into a network that encircles the earth. From a little chateau in Europe to Indiana University, miles of wire and fiber optics are connected. A new communications standard is born. Computer to computer to computer....

The Internet surrounds our lives; a "web" of knowledge so all-encompassing that you may never see it all in 20 lifetimes. Individuals, private industry, large conglomerates, universities, schools, libraries and governments all over the world are now part of the Internet world. It is used for databases, libraries, advertising, quick communication, research, entertainment, ad infinitum, and more. Included in this vast database, of course, is electronics.

Without the growth of electronics, the Net would not exist. So it is very fitting that electrical subjects, datasheets, parts houses and manufacturers are so easily found in the Net. The Internet replaces a wall full of datasheets, a library full of electronic reference books, hours of phone calls, and of course, expensive postage. The Information Superhighway: where do you find it, and how do you search for it?

On the Internet there are locations for stored data; these are called *sites*, or *URLs* in Net language — Universal Resource Locators. Think in terms of a tourist site; go buy a map and look for locations of interest. You may relate the street names or landmarks to URLs. Microsoft's headquarters are located in Seattle; on the Net, they are located at "www.microsoft.com." URLs will be explained in detail later.

This book provides you with two types of information. First of all, this book will show you how to expand your information and entertainment-hunting abilities. You will be able to survey the electronics industry through the "eyes" of your computer — find valuable information in a flash; know what sites are worth 5 minutes of your time and which are worth an hour; learn how to do searches; find information on manufactures, distributors, retail outlets, hobbyists, project resources, Email addresses, etc. This text acts as a cyberspace road map for the subject of electronics, allowing you to navigate the I-way with ease. The second major portion of this book contains a listing of sites for you to browse through, so you can see what the electronics industry has to offer over the Internet — everything from army missile operating systems to a Zen theory of electronics.

Please keep in mind that the Internet is increasing in size at an exponential rate, with new sites coming online daily — hundreds, if not thousands. Learning to become a competent I-searcher will enable you to zip right to the data or program you need.

The Internet and this directory provide a mere starting point to allow you to branch out. At the Net's base is data, and at its apex is discovery. The sum of human knowledge is out there for you to discover, and easily accessible through the Net. Now it is your turn to make use of it — to venture into this new-found electronics frontier.

Surf's up!

INTERNET 101

Learning the Internet is now a manageable task. Electronic hobbyists and professionals will find the newest cyberspace software simple compared to ancient programs. A sense of exploration will develop as your familiarity rises. Your intuitive mind will take over and soon you will be caught in the Net: "What's that? Let's see where this path takes me. What else can I find on the Net? How much time do I have left on this thing? Can I just stay on longer, please?" A few lessons, a few mouse clicks, and soon you will be making use of this new electronic tool.

WHAT IS THE INTERNET?

The Net, as it is commonly called, is a network of approximately 35-40 million computers linked together through communication lines. It's that simple. The computers can be any device from a PC in your living room to a modern supercomputer at DEC. Each, in turn, can communicate a common language and transfer data back and forth across the earth.

HOW CAN THE INTERNET HELP AN ELECTRONICS HOBBYIST/PROFESSIONAL?

Knowledge is a tool just like your multimeter or oscilloscope. Stored in computers among the Internet are mass amounts of data. Trillions of bytes of knowledge. This information can be anything from a program that calculates Ohm's law to a schematic of a satellite system. Between searching for the sports news and stock quotes, electronics oriented individuals can locate that elusive part for their ailing VCR. A circuit designer can find a schematic routing program which will solve a design flaw in his latest endeavor. A hobbyist can locate a new project to keep his soldering iron busy. A corporate buyer can get the latest prices on microprocessors. You name the piece of electronics data you need; IT'S ON THE NET!

Another factor that may aid a hobbyist or professional is the ability to communicate near-instantaneously to anyone around the world. Millions of electronics oriented individuals are on the Net and want to keep in touch with others in their field. If you don't know the answer or can't find it yourself, someone else can help. All it takes is an Email address and a computer instead of a paper, pen and stamps.

HOW MANY PEOPLE ARE ON THE NET?

There are no proven statistics, but current estimates state that there are 35-40 million Internet users. The Internet is HOT and people are signing up in droves. It is estimated that by the year 2000 there will be 100 million computers and up to 200 million users comprising the Net.

WHY ARE SO MANY PEOPLE TALKING ABOUT THE NET?

The Net has been in service, as we know it, since the 1970s. What has made it hot in recent times is a famous speech by Al Gore quoting the Information Superhighway. With typical political runaround, it was not clearly defined what that term meant, but we all wanted it! The Internet seemed to fit the "bill," and the media blitz connected Gore's speech with the fast-growing Internet.

WHAT IS AVAILABLE TO VIEW ON THE INTERNET?

A brief summary of available resources on the Net:

1. Books, magazines.
2. Chat lines.
3. Company contacts.

4. Company information.
5. Contests.
6. Email.
7. Firmware.
8. Free product samples.
9. Frequently Asked Questions (FAQs)
10. Hardware.
11. IC pinouts.
12. Newsgroups (discussion groups) relating to all sorts of electronics topics.
13. Other hobbyists/technicians/engineers to converse with.
14. Product applications.
15. Product datasheets.
16. Product information.
17. Product pictures.
18. Product support.
19. Product support software.
20. Product wholesale/retail outlets and distributors.
21. Product wholesale/retail pricing.
22. Projects.
23. Software.
24. Technical support.

See the chapter *Internet Resources* for a more comprehensive list.

WHAT YOU NEED TO GET ONLINE

A modern PC will do just fine. Most electronics individuals may find they already have the minimum hardware outlay buried in their workshop or closet. You don't need the latest Pentium to get online. An old trash computer can be turned into a treasure-trove with the addition of a modem.

There is a wide range between the low-end and the high-end computer system. Find the hardware that fits your budget.

A Low to Mid-Range System
1. IBM compatible 386 or Mac equivalent. An old 286 can even be used with a few cheap upgrades.

Figure 1. Hobbyist electronics sites are great sources for information. Sam J. Engstrom's site has a CMOS Functional Diagrams page to help you save search time.

2. 4 Megs of RAM minimum. Eight is more realistic. RAM is obscenely cheap now.
3. 14.4 modem minimum. Don't waste your time on anything smaller. These modems are cheap.
4. DOS 5, Windows 3.11 or Mac OS.
5. A Slip or PPP account from your local Internet Service Provider (ISP). Shop around.

6. Software. Don't run out to a computer shop and layout $50 for software. This book contains sites for you to download the necessary programs on a shareware/ freeware basis. Some providers will even offer registered programs free when you sign up with them.

A Mid-Range to High-End System

1. IBM compatible 486 to Pentium or Mac equivalent.
2. 16-32 Megs of Ram.
3. 28.8-ISDN Modem.
4. Windows 95/NT, Linux or Mac OS.
5. Registered software.

Don't be embarrassed to pull out that old 286 and mono-chrome monitor. A supercomputer isn't necessary to surf the Net. The first computer I used for access was a 386DX40 with only 4 megs of RAM and a 14.4 modem. Coming soon, next to the video game consoles, will be a black box that will allow you to surf the Web and send Email right from your TV. The $250-$500 price tag may stop you from spending $2500 for a Pentium, but the device is in no capacity a PC.

HOW MUCH IS IT?

The Internet has to be the best deal of this century. Where else can you send mail, experience loads of entertainment and locate information from the largest databases in the world, all for only $15 per month?

The cost of Internet accounts (PPP or SLIP) vary by a wide margin. One dollar per hour is average but in big cities $15-$30 a month for unlimited time is typical. A Shell account runs $8-15 a month but is difficult to operate for the untrained user. Ask your friends which service they are happy with. Shop around and find the best deal.

HOW TO ACCESS THE NET TO FIND ELECTRONICS DATA IF YOU DON'T HAVE A COMPUTER

If you don't have a PC, you may find access at your local university or library, although your time may be very limited.

Many electronics stores across the country are taking advantage of the Internet by offering "Internet Resource Centers" on their premises. You can download datasheets, look at company sites on the WWW, and most important, get advice from Internet professionals. Take this book with you and ask to use their computer to surf.

HOW TO FIND WHAT YOU ARE LOOKING FOR

Finding information on the Internet can be a grueling chore; the Internet is infamous for it. This is partially true, but you must consider the amount of data being searched. Once you see the number of documents the Net is dealing with, your search time will be put into perspective.

The *Web Directory* section in this book gives you a convenient way to zip to electronics information. If the site you need is not listed, you must then do an Internet search. New devices are coming out that allow very descriptive variables to help you narrow your search. Find a method that is comfortable. Later in this book is a section titled *Teaching Yourself to Fish*, which will better prepare you.

IS IT DIFFICULT TO LEARN THE INTERNET?

It is to a degree; but nowhere near as difficult as learning such computer endeavors as Microprocessor Assembly Language. Once a few programs are under your belt, the rest is child's play.

WHAT ARE THE TERMS OF THE INTERNET?

This book will explain the common terms of the Net as you go along. If you have trouble finding a definition, go to the *Glossary* or find a simple Internet book at the library. There will be a wall full of them.

DO YOU REALLY NEED THE INTERNET?

Before calling a provider, you may want to ask yourself:

1. What do I want the Internet for?
2. Will it give me what I want?
3. How can I make it fit my budget?
4. Can I make use of the Internet for other things?
5. Will I be left behind in the future if I do not get online now?

Are you convinced the Internet is just the right tool to add to your workbench? Then get ready to become part of the future of electronics.

INTERNET RESOURCES

The Internet is an ever-changing entity that in its infancy was a modest research/military tool.   With time and technology, the Internet propagated unceasingly. It grew from a text based research tool into a way to communicate simple messages. Soon after discussion groups of all forms bloomed. Tim Berners-Lee of CERN, in 1989, then became a stepfather to this child and created its greatest accomplishment to date: the World Wide Web, with its sights, sounds, and links to millions of computers around the globe.

This new Internet now spoke a new language: Netese. "Web-surfing," "Spamming," "Flaming," "Net-Cops," and "Cyberspace" are a few of its words. Millions wanted to learn it and let their computer skills grow with it. Thus, the Net as we know it was born.

This new, grown-up Internet now has resources beyond belief. New Internet applications pop up daily. Some become well nested in cyberspace and others are for practicing your file deleting skills.

WHAT ARE THE ELECTRONICS-RELATED INTERNET RESOURCES?

What resources will the average hobbyist, technician or engineer use? How do I access the electronics information I want on the Net? The following are tried-and-true resources that should be of great interest to you. This is by no means a complete list; but, it will give you a basic overview and give you something to do between burning your fingertips on that soldering iron or getting yelled at by your boss:

1. ARCHIE
2. EMAIL
3. FAQs
4. FTP
5. GOPHER
6. IRC
7. LISTSERVs
8. NEWSGROUPS
9. PDF Files
10. TELNET
11. WWW

EMAIL

Each person on the Internet has a personal Email (or *electronic mail*) address. To communicate with another person, you simply open up a program on your computer, type your message, and out it flies at light speed to the recipient. Federal Express has nothing on the speed of Email. It's a fast, simple, and powerful tool for any company or individual.

WHAT CAN YOU SEND WITH EMAIL?

With the Email programs in use today, you can send and receive text messages and files; but promises of new program features are always outstanding.

WHAT DOES THE "@" SYMBOL MEAN, AND WHAT IS YOUR ADDRESS?

When you sign up with an Internet service provider, you are given a user ID. The ID is the first part of your address. The "@" simply means "at." The text that follows the @ is the server that your account uses.

EXAMPLE: **joe@here.com** would be **Joe** at the service provider called **here**. By the way, the "." is a pronounced dot, and *com* means that the provider is a commercial server. See the *Glossary* for additional domain suffixes.

HOW YOU CAN USE EMAIL

Some companies prefer that you send Email rather than call them. They're then able to handle communications at their own convenience. It sure beats being on hold for 1/2 hour on a long-distance call. If you want to send someone a file, attach it to an Email message. A robot (of sorts) called a *Listserv* can send you Email automatically. Soon, most communications probably will be sent via an electronics medium, so you should earn to utilize Email now.

WHERE TO GET PROGRAMS FOR EMAIL

PC Eudora, Pegasus, PINE, ElmPC, and many other mail programs are available through *Tucows* at HTTP://tucows/ softmail.html, or *CDROM.com* at HTTP://www.cdrom.com/ pub/simtelnet/.

Netscape Navigator and Microsoft Internet Explorer also have built-in (limited) Email programs. See the *World Wide Web* chapter for a list of addresses.

My personal choice is PC Eudora. It offers advanced features and comes as postcardware (free if you send the company a postcard), shareware, and registered versions. If you have a DOS or UNIX-only computer, consider ElmPC or PINE, available at CDROM.com.

HOW TO FIND SOMEONE'S EMAIL ADDRESS

Look up HTTP://www.qucis.queensu.ca/FAQs/email/

HOW TO FIND OUT MORE ABOUT EMAIL

HTTP://www.cis.ohio-state.edu/hypertext/faq/usenet/mail/ top.html

FAQs — FREQUENTLY ASKED QUESTIONS

A FAQ is a document which contains commonly-asked questions and answers about a given topic. These files are placed in Web pages, in newsgroups, on IRC, etc. A FAQ is simply a universal way to educated people about the Net in areas of unfamiliarity; for example, this book is modeled after a FAQ. FAQs are a natural way to learn as opposed to standard methods, and seem to have been fully adopted by "Net citizens."

WHAT QUESTIONS DO FAQS ASK?

FAQs can ask *any* question:

"How do you measure voltage in a circuit?"
"What is this newsgroup and what can I post on it?"
"Where can I find out more about this?"
"Where do I find a FAQ on FAQs?"

(PS — HTTP://www.jazzie.com/ii/internet/faqs.html will answer the last question).

HINT

When in doubt about a resource or newsgroup, look for a FAQ that contains the information you need. Use the Web and use the keyword "FAQ" in your search. You may also want to look for Primers on those subjects.

FILE TRANSFER PROTOCOL

File Transfer Protocol (FTP) is the protocol that allows you to upload or download files from the Internet. With this resource, companies and individuals can post electronics-related files for you to access; done using a program which handles the transfer. Here are a few terms, software locations, and a place to find out more about FTP:

Anonymous Login: In order to download or upload files, you must log into an FTP server (computer) with a password. Using *anonymous* as your user ID will allow limited server access — otherwise, you need a valid user ID and password.

Directory(ies): The file hierarchy for the FTP server.

Download: The act of moving files from the FTP server to your computer.

FTP Server: The computer on the Net that contains the files.

Index.txt: The text file that has a description of the files in a directory. Some programs will automatically access this for you so you can view the hierarchy and files.

Upload: The act of moving a file from your computer to the FTP server.

Several FTP programs are available via the Web or the FTP itself. Find one that matches your taste. Most sites are now accessible through the Web, but I still use CuteFTP, more often than not, when I am in a hurry and don't want to wait for Netscape to load into RAM.

Cute FTP: A small, simple program good for computer systems with low resources: HTTP://www.cuteftp.com/; FTP:// ftp.cuteftp.com/

Microsoft Internet Explorer: You can access FTP servers through this program by placing an FTP:// then the address in the URL window. Example: FTP://ftp.microsoft.com/. See the *World Wide Web* chapter for addresses.

Netscape Navigator: You can access FTP servers through this program as well; same way as Internet Explorer. See the *World Wide Web* chapter for addresses.

WS_FTP: Another, simple Winsock FTP program. See HTTP://www.tucows.com/ for the current version.

HOW TO FIND MORE INFORMATION ABOUT FTP

HTTP://www.dnet.net/FAQs/ftp-faq.html

INTERNET RELAY CHAT

Email and USENET postings are slow to receive responses; having to wait days or weeks is typical. Enter the IRC, or Internet Relay Chat. The IRC lets you communicate with others in REAL-TIME; no waiting. You can type someone a message on your screen, and as fast as the recipient can type, you can have an answer.

The IRC is currently an under-used resource in the field of electronics, which is a shame. Two-way communication is a valuable commodity these days. As more people learn about the IRC and become used to it, it will rise in popularity. What better way to get answers to your questions RIGHT AWAY?

IRC TERMS

Channel: A station (of sorts) where people converse about the topic at hand. Example: #electronics.

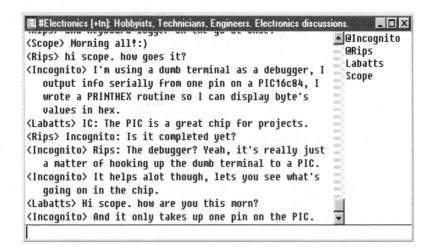

Figure 1. Internet Relay Chat can be used for real-time conversations.

Channel Operator: The person who started and now controls the channel, or someone who was given the status by the originator. Look for the @ sign before their name. Example: @Nick.

IRC Server: A computer on the IRC network. You can login through your service provider in order to talk to others on the network. There are IRC servers for different networks, such as DALnet, EFNet, Undernet, etc. However, people on one network cannot talk to people on another network. EFNet is by far the most popular network for electronics-related information.

List: This is the list of channels that are currently open. The list can contain as many as 6,000 channels.

Nickname: The name you choose to call yourself while on the IRC. Example: Rips (the author).

Private Message: A way to talk privately with someone on a one-to-one basis, outside of a channel.

Topic: The conversion currently being discussed in the channel. Sometimes the topic is listed as simple gibberish.

Symbols and Acronyms:
:) or :-) is a smile. Look at it sideways.
:(or :-(is a frown.
:-O expresses shock, as in "Oh, my!"
:-P is a tongue sticking out.
BRB means *Be Right Back.*
BTW means *By The Way.*
LOL means *Laughing Out Loud.*
ROFL means *Rolling On Floor Laughing.*

WHAT CHANNELS ARE OUT THERE FOR ELECTRONICS AND RELATED TOPICS?

There are limited channels for electronics, but only because few people have opened them up. You can start a channel called "#vcr_repair" as easily as typing /join #vcr_repair. You would then have the status of Operator of the channel for the duration of your stay. So, if you want more electronics-related channels, go ahead and create them, and see who shows up.

Here is a list of electronics and related channels to help you. Explore one to see current information and the number of people involved:

#Amiga
#Atari
#C
#Calc-TI
#CarStereo
#Circuits
#Computer
#Computer_help
#Electronics
#Engineering
#GameCode
#Gamers
#Ham-Radio
#Intel
#Laser
#LaserService
#Math
#Mathematics
#Physics
#Pro-Audio

NOTE

EFNet channel **#Electronics** is where you will find almost anything relating to electronics. Join the channel and ask the operators for more information, or visit: HTTP://pobox.com/~electronics/irc/

WHERE TO FIND MORE INFORMATION AND PROGRAMS FOR IRC

Internet Relay Chat FAQ: HTTP://www.kei.com/irc.html

For a list of current IRC programs, search Tucows, Shareware.com, or see the IRC FAQ: HTTP:// www.tucows.com/ or HTTP://www.shareware.com/

PDF — PORTABLE DOCUMENT FORMAT

Many files on the Internet are stored as PDF files. These would include datasheets, magazines, catalogs, schematics, etc. It gives the document creator the ability to encode several types of documents, such as spreadsheets, word processing documents, databases, graphics, etc. It also enables the document to be read with several different platforms, printed in a professional, camera-ready manner.

To view PDF documents, you currently need a program from Adobe Systems Incorporated. In the future, Web-browsers will have PDF programs built-in. You can obtain a copy of Adobe Acrobat at HTTP://www.adobe.com/acrobat/ readstep.html.

You will need to add Adobe to your Internet resource programs list if you will be viewing any datasheets from the Web.

TIP
When printing a PDF file choose the Shrink to fit print option to ensure the entire area of the page is reproduced.

ADDITIONAL INTERNET RESOURCES

These resources are still in use, but have been greatly overshadowed by the preceding ones. With time and technology, even the Web may be relegated to this list; no one knows what the future will bring or what it will contribute to hardware and software.

ARCHIE: A resource that allows you to search the Net for computer files that have a specific file name. This is quite a slow and inaccurate method. The Web has all but replaced this resource; but, if you know exactly what you are looking for, this program will show you all of the sites that contain it.

BITNET: A global network for academic institutes and research organizations. It is separate from the Internet.

GOPHER: The precursor to the World Wide Web. It has one major drawback: no graphics. However, if you need raw text data, this program is the only way to fly.

LISTSERVs: An automated mailing list that will send data to your EMAIL address periodically. The list of LISTSERVs grows as subjects grow.

TELNET: An older Internet resource that gives you remote access to computers on the Net so you don't have to call up the computer's specific phone line. This archaic service is relegated to mostly government and public libraries that require its use.

NEWSGROUPS

They have been called "overloads of opinions" and "word wars," but on the whole they are an invaluable source of information. I am referring to the USENET; commonly known as newsgroups. This Internet resource mimics bulletin boards on library walls; ads, addresses, announcements, discussions, letters, notes and opinions are pasted all over it.

The supply of electronics-related information in these newsgroups is unlimited. What information can be found under these electronic "push pins"? What do you need to know to make use of this resource?

WHAT ARE NEWSGROUPS?

The USENET is one huge bulletin board broken down into sections called newsgroups. Each newsgroup is then broken down into *articles* and *threads*. Articles are postings made by individuals, and threads are responses to an article.

Newsgroups are a global way to communicate your ideas or to read the views of others. Approximately 15,000 to 20,000 groups are currently operating, with topics ranging from alien spacecraft to zoology. In between are groups useful to the electronics hobbyist and professional. Electronics-related groups are as far-ranging as any subject, and this book provides a list to help you find a newsgroup(s) that fulfills your needs.

WHAT'S ON THE USENET FOR ELECTRONICS?

Newsgroups may contain any or all of the following:

1. Addresses for sites on the WWW or FTP.
2. Advertisements.
3. Articles relating to subjects at hand.
4. ASCII schematics.
5. Binary files.
6. Classifieds.
7. Discussions.
8. Executable files.
9. MPEG movies.
10. News/Press releases.
11. Opinions.
12. Personals.
13. Pictures.
14. Product information.
15. Programs.
16. Requests for help.
17. Requests for information.
18. Sounds.
19. Want ads.

On occasion you will find executable programs, pictures, sounds, etc., in a newsgroup. The use of text files is three-fold — you can communicate using text-based information, illustrate ideas using ASCII characters, or direct people to an FTP, WWW site, or other URLs where further information is available.

EXAMPLE TEXT

>> Can anyone tell me how a circuit works?

EXAMPLE ILLUSTRATION

```
>>        —| 2k |—
>> |                    |
>> — +                  |
>> -    a simple        |
>> —      circuit       |
>> - -                  |
>> |12V                 |
>>     ----------------
```

EXAMPLE DIRECTION TO URL

>> Please visit our newest circuit basics
information site at http://pobox.com/~elec-
tronics/basics/

WHAT YOU NEED TO VIEW NEWSGROUPS

Besides the requirements listed in the chapter *Internet 101*,
you will need a program called a *newsreader* so you can view
the USENET. Here is where to procure a shareware or
freeware newsreader:

1. Netscape Navigator has a built-in newsreader that is
   quite simple to operate. It can be obtained from:
   HTTP://home.netscape.com/
   or FTP://ftp1.netscape.com/

2. Microsoft Internet Explorer also has a built-in
   newsreader. It can be obtained from:
   HTTP://www.microsoft.com/ie/ie.htm

Also available are News Xpress (NX), Free Agent, and
WinNV at HTTP://tucows/softnews.html.

Your final option is to check Shareware.com for the latest
programs, at HTTP://www.shareware.com/.

Try a few newsreaders and see which suits your needs. Make sure your system can handle the software efficiently — i.e., don't run Netscape 3.0 on an old 386 with 4 megs. My personal choice is a combination of Netscape and News Xpress. They complement each others' weak and strong points.

HOW TO USE A NEWSREADER

Find the HELP or README file that accompanies the viewer and learn the commands well. Here are some descriptions of options to help you:

Show All Newsgroups: Displays all newsgroups on your Internet provider's server. Not all newsgroups may be on your provider's server. Contact your provider for further information.

Subscribe To: Lets you download, on a regular basis, the newsgroups you prefer.

Post Article: Posts stand-alone articles to a newsgroup.

Post Reply: Creates a thread by posting follow-ups to someone's article.

Mail Reply: Sends an Email message to the author of an article.

IS THERE A WAY TO DO ONLINE SEARCHES FOR NEWSGROUPS OR ARTICLES?

The following WWW-based search engines contain search tools to look for newsgroup articles of interest. You can search by date, newsgroup or word:

Alta Vista @ HTTP://www.altavista.digital.com/

Awebs Newsgroup Archives @ HTTP://awebs.com/ news_archive

DejaNews @ HTTP://www.dejanews.com/

Excite @ HTTP://www.excite.com/

Infoseek @HTTP://guide.infoseek.com/

Reference.com @ HTTP://www.reference.com/

The Usenet Newsstand @ HTTP://criticalmass.com/concord

Check out Netscape's site for further information on these search engines' capabilities: HTTP://home.netscape.com/escapes/search/usenet.html

To learn search techniques, read the chapter *Teaching Yourself To Fish*.

TIP
If you use Netscape or Microsoft Internet Explorer browsers while surfing, there are links to newsgroups that can automatically open up your newsreader and the selected group. Take advantage of this short cut.

DOES IT COST ANYTHING TO SUBSCRIBE TO NEWSGROUPS?

I am asked this valid question frequently by Internet students. There is no cost to read or post to a newsgroup, save the monthly cost you pay to your Internet provider.

HOW TO APPROACH NEWSGROUPS

1. Read these UseNet primers:

HTTP://www.tezcat.com/~abbyfg/faq/what-is-usenet.html
HTTP://www.crpht.lu:10007/FAQ/usenet/primer/part1/part1.html

Or do a search with the keywords "USENET" and "Primer."

2. Survey the newsgroup for a few days. Find a FAQ (Frequently Asked Questions) about the newsgroup and read it. There is usually one posted continually to each group, or check out:

HTTP://www.cis.ohio-state.edu/hypertext/faq/usenet/FAQ-List.html
HTTP://hamsterix.funet.fi/index/FAQ/usenet-by-hierarchy/

Once this is done you'll be on steadier ground.

3. Decide who the flamers are and treat them like they have the plague. Never respond to or bother reading their rantings unless they're well-founded.

4. Don't let the word 'newbie' put you off. We were all new to the Net at one time.

5. Find a group that is comfortable for you. If you feel uncomfortable with the people posting in it, find another.

6. Post a few article responses once you are confident. If you are still uncomfortable, Email the people first with your responses.

7. Post a few stand-alone articles and see what discussions you can create in the group.

8. Utilize the Usenet for your hobby or professional pursuits.

9. Have fun! It's only as serious as you want it to be. :-)

WHAT ARE FLAMING AND SPAMMING?

Flame n. : An insulting message.

Flame, Flaming v. : To send someone an insulting message or post an insulting response to their article.

Example:
Someone posts:
```
>> I believe this new chip is a wonderful
addition to Acme's line of products.
```

Reply or "Flame":

```
>> Who asked your opinion?
```

This would be a tame flame; most are not nearly as kind.

Spam n: A message, usually some kind of advertisement, posted over multiple newsgroups, that has nothing to do with the newsgroup at hand; off-topic messages.

Spam, Spamming v: The act of placing spam on the USENET.

Example:
The sci.electronics newsgroup has this message posted:

```
>> Make a million dollars in 30 days with
this simple  money making program!
```

It has nothing whatsoever to do with electronics, so why is it there?

NOTE

Please don't put off posting your views to a newsgroup because you are expecting to be flamed. Just be sure your posting relates to the subject at hand. Unfortunately, as the adage says, "Opinions are like...". Keep in mind that newsgroups are full of opinions. Don't let this stop you. There are a limited numbers of flamers out there. If all else fails, remember the other old adage, "Sticks and stones..."

Be aware that newsgroups exist for posting your messages. Even the million-dollar message above has its place. Put a design question in *sci.electronics.design* and not in *sci.electronics.basics*. Seek out the right newsgroups, and you will never have to worry about being "flamed for spamming." At least not from me.

HINTS TO KEEP THE USENET FREE FROM GARBAGE

1. NEVER reply to a flame. If you find you must, please remember you are only clogging up the group each time you do; so always be polite and explain why you have posted that information to that group. Then leave it at that since the majority of readers will side with you.

2. Learn what the newsgroup's topic is and only post topics relating to that subject. There are plenty of other newsgroups for on-topic opinions and articles. People will happily read and respond to your on-topic postings.

TIP

If someone else has posted a reply to a question, don't post the same thing or post an "I agree" message. It may only be a few characters, but by the time you add all of the attached information, the posting will be 10-100 times its original size. Useless bytes equal crowded newsgroups, which equal slow download times.

WHAT DO ALT, COMP AND SCI MEAN?

These are the prefixes for electronics newsgroup topics:

alt: Means *alternative*. Alt newsgroups can be started by anyone and thus are usually the most wild, unmoderated groups. However, they are still a great forum for discussions.

biz: Business-oriented topics.

comp: Computer-oriented topics.

ieee: Institute of Electrical and Electronics Engineers-related topics.

misc: Topics that don't fit anywhere else.

rec: Recreational-oriented topics.

sci: Science-oriented topics, including electronics

Examples:
```
alt.computer — biz.comp.hardware —
comp.hardware — ieee.announce —
misc.industry.electronics.marketplace —
rec.radio.amateur — sci.electronics
```

HOW TO USE NEWSGROUPS TO HELP YOUR HOBBY/ PROFESSION

The evolution of UseNet continues. Electronics is a well-covered area, and will evolve with the rest of the Internet topics. Right now, it is best to roll with it and gather whatever articles interest you. You should ask questions that you are having trouble answering, post announcements of your products, and tell people your theories. Tell people about projects or research you are working on; or get your employees into groups that relate to your business and have them help Netusers. Voice your opinions, disseminate valuable information, or start a discussion on topics that interest you. The only limit to this resource is self-imposed, so eliminate your doubts. After all, Newsgroups are a forum to exercise the right of free speech while helping other hobbyists and professionals in the Net community.

NEWSGROUPS DIRECTORY

The following is a list of USENET newsgroups that were operational when this book was compiled. Some of these newsgroups may not be accessible through your Internet provider's server. If missing, you may be able to contact your ISP and request they post that newsgroup for viewing. To find new groups relating to electronics:

1. Use the SHOW ALL NEWSGROUPS feature to get a complete list.

2. Once you have all newsgroups on your computer, use the SHOW NEW NEWSGROUPS feature. This will give you a listing of the newest newsgroups.

3. See HTTP://pobox.com/~electronics/ for the authors updates.

4. Keep an eye out for links to newsgroups on Web pages.

NOTE: Because each newsgroup continually evolves it's "purpose," I will leave it to the LIST OF USENET FAQs to explain each group:

HTTP://www.cis.ohio-state.edu/hypertext/faq/usenet/FAQ-List.html
HTTP://hamsterix.funet.fi/index/FAQ/usenet-by-hierarchy/

Here is a brief explanation of the **sci.electronics** newsgroups at the time of this writing:

sci.electronics	Circuit, theory, electrons and discussions.
sci.electronics.basics	Elementary questions about electronics.
sci.electronics.cad	Schematic drafting, PCB layout, simulation.
sci.electronics.components	ICs, resistors, caps.
sci.electronics.design	Electronic circuit design.
sci.electronics.equipment	Test, lab, and industrial products.
sci.electronics.misc	General discussion of the field of electronics.
sci.electronics.repair	Fixing electronics equipment.
misc.industry.electronics. marketplace	Electronics products and services.

DIRECTORY

alt.2600
alt.amateur-comp
alt.binaries.emulators.cbm
alt.books.technical
alt.cellular-phone-tech
alt.comp.compression
alt.comp.emulators.executor
alt.comp.freeware
alt.comp.hardware.homebuilt
alt.comp.hardware.homedesigned
alt.comp.pc-homebuilt
alt.comp.periphs.mainboard.asus
alt.comp.periphs.mainboard.tyan
alt.comp.shareware
alt.computer
alt.computer.drivers.wanted
alt.computer.workshop.live
alt.control-theory
alt.design.product
alt.electronics.analog.vlsi
alt.electronics.manufacture.
    circuitboard

alt.emulators.ibmpc.apple2
alt.engineering.electrical
alt.engr.dynamics
alt.future.millennium
alt.guitar.amps
alt.how-to.create.a.newsgroup
alt.illustration.technical
alt.info-science
alt.intel
alt.machines.misc
alt.mag.nuts-volts
alt.microcontrollers.8bit
alt.online-services.compuserve
alt.online-services.delphi
alt.online-services.genie
alt.online-services.microsoft
alt.online-services.prodigy
alt.radio.amateur.club-clarc
alt.radio.cb.crunch
alt.radio.digital
alt.radio.internet
alt.radio.scanner

alt.radio.scanner.uk
alt.satellite.tv.crypt
alt.technology.misc
asu.comp.micro
aus.computers.ai
aus.computers.logic-prog
aus.computers.parallel
aus.electronics
aus.radio.amateur.digital
aus.radio.amateur.misc
biz.books.technical
biz.comp.hardware
biz.marketplace.computers.
     discussion
can.schoolnet.elecsys.jr
can.schoolnet.elecsys.sr
can.vlsi
comp.ai.doc-analysis.misc
comp.ai.doc-analysis.ocr
comp.ai.edu
comp.ai.fuzzy
comp.ai.neural-nets
comp.ai.philosophy
comp.answers
comp.arch
comp.arch.bus.vmebus
comp.arch.embedded
comp.arch.fpga
comp.arch.storage
comp.dsp
comp.editors
comp.emulators.misc
comp.hardware
comp.home.automation
comp.hypercube
comp.ibm.pc.hardware
comp.lang.(*.*)
comp.lsi
comp.lsi.cad
comp.lsi.testing
comp.org.ieee

comp.patents
comp.periphs
comp.programming
comp.protocols.misc
comp.publish.electronics.
     developer
comp.publish.electronics.
     end-user
comp.publish.electronics.misc
comp.realtime
comp.robotics
comp.robotics.misc
comp.robotics.research
comp.software
comp.speech
comp.speech.eng
comp.std.wireless
comp.sw.components
comp.sys.dec.micro
comp.sys.harris
comp.sys.hp.hardware
comp.sys.ibm
comp.sys.ibm-pc
comp.sys.ibm.pc.hardware
comp.sys.ibm.pc.hardware.chips
comp.sys.ibm.pc.hardware.comm
comp.sys.ibm.pc.hardware.misc
comp.sys.ibm.pc.hardware.networking
comp.sys.ibm.pc.hardware.storage
comp.sys.ibm.pc.hardware.systems
comp.sys.ibm.pc.hardware.video
comp.sys.ibm.pc.misc
comp.sys.ibm.pc.programmer
comp.sys.ibm.pc.soundcard.tech
comp.sys.intel
comp.sys.mac
comp.sys.mac.comm
comp.sys.mac.databases
comp.sys.mac.hardware
comp.sys.mac.hardware.storage
comp.sys.mac.hardware.video

comp.sys.mac.hardwarecomp.
    sys.mac.comm
comp.sys.mac.misc
comp.sys.mac.programmers.(*.*)
comp.sys.mac.system
comp.sys.pc.hardware.misc
comp.sys.sgi.hardware
comp.sys.sun.hardware
comp.unix.(*.*)
comp.windows.(*.*)
cu.vlsi
de.alt.binaries.pictures.tech
de.sci.electronics
fj.comp.dsp
fj.engr.elec
fj.enjr.robotics
fj.org.ieee
fnet.hypercubes
francom.electronique
git.tech.futures
ieee
ieee.announce
ieee.bbs.help
ieee.bbs.info
ieee.ces.audio-visual
ieee.ces.home-automation
ieee.ces.personal-
communications
ieee.config
ieee.eab.announce
ieee.eab.general
ieee.general
ieee.pcnfs
ieee.pcs.general
ieee.pub.announce
ieee.pub.general
ieee.rab.announce
ieee.rab.general
ieee.region1
ieee.stds.announce

ieee.stds.general
ieee.students
ieee.tab.announce
ieee.tab.general
ieee.tcos
ieee.usab.announce
ieee.usab.general
in.ham-radio
intel.network
mcgill.general
misc.books.technical
misc.industry.electronics.marketplace
nctu.ee.general
nctu.ee.vlsi-cad
rec.antiques.radio+phono
rec.arts.tv.interactive
rec.ham-radio
rec.ham-radio.packet
rec.ham-radio.swap
rec.radio.amateur
rec.radio.amateur.antenna
rec.radio.amateur.digital.misc
rec.radio.amateur.equipment
rec.radio.amateur.homebrew
rec.radio.amateur.misc
rec.radio.amateur.packet
rec.radio.amateur.policy
rec.radio.amateur.space
rec.radio.amateur.swap
rec.radio.broadcasting
rec.radio.cb
rec.radio.scanner
rec.radio.shortwave
rec.radio.swap
rec.video.satellite.dbs
rec.video.satellite.europe
rec.video.satellite.misc
rec.video.satellite.tvro
relcom.commerce.computers.
    ctrlsystems

relcom.commmerce.computers.
      ctrlsystems
sci.electrical.writing
sci.electronics
sci.electronics.basics
sci.electronics.cad
sci.electronics.components
sci.electronics.design
sci.electronics.equipment
sci.electronics.misc
sci.electronics.repair
sci.engr.control
sci.engr.manufacturing
sci.engr.semiconductors
sci.engr.television.advanced
sci.engr.television.broadcast
sci.homebrew
sci.logic
sci.nanotech
sci.optics
sci.optics.fiber
sci.physics.electromag
sci.polymers
sci.research
sci.space.tech
sci.techniques.mag-resonance
slac.rec.ham_radio
tamu.electronics.library.resources
tnn.radio.amateur
triangle.vlsi
tw.bbs.sci.electronics
umn.ee.(*.*)
uw.ieee
uw.neural-nets
uw.vlsi
uw.vlsi.software
uw.vlsi.system
uwarick.societies.amateur-radio

THE WORLD WIDE WEB

If your curiosity for the Internet has compelled you to purchase a computer and an online account, it is likely due to the hype and hoopla, entertainment and silliness, commercialization and information of the World Wide Web. Fast growth in advertising and entertainment launched the Web into a cult-culture icon. With an estimated 60 million web pages online you will never be deprived of electronics projects and resources to skim through.

Internet worlds are inexpensive to create. The World Wide Web opens up new vistas in information dissemination. In the new landscape lie programs, pictures, circuits and contacts to converse with real human beings. Millions travel these roads and are mesmerized by its natural user-friendly format. Sucked in by its appeal and information, they never know what hit them.

WHAT EXACTLY IS THE WORLD WIDE WEB (WWW)?

1. The Web is a database of documents (web pages) currently numbered in the millions. They are stored on hundreds of thousands of computers around the world in the form of *hypertext* documents. Each user accesses files with a program that converts hypertext into text, pictures and sounds, and displays them on their computer screen to enjoy.

2. The WWW is the most important resource to come along since the multimeter.

HOW THE WEB CAN HELP YOU, AS AN ELECTRONICS HOBBYIST/PROFESSIONAL

Here is a list of common electronics resources that reside on the WEB.

1. Animations
2. Application sheets for components/products
3. Articles
4. Books online
5. Cameras displaying pictures from around the world
6. Company information
7. Computer resources
8. Contacts
9. Contests
10. Databases
11. Data sheets for almost any electronics component/product
12. Drivers
13. Educational courses
14. Email addresses
15. Employment opportunities
16. Engineering resources
17. Help with products
18. Hints
19. Hobby resources
20. Inventories
21. Libraries
22. Linecards
23. Links to other similar resources around the Web
24. Lists
25. Magazines online
26. Movies
27. Newspapers
28. OEM
29. Online shopping/ordering
30. Other hobbyists/professionals
31. Papers
32. Parts/component locators
33. Pictures
34. Product explanations
35. Product reviews

36. Product specifications
37. Programs
38. Projects
39. Repair resources
40. Research
41. Samples ordering
42. Schematics
43. Schools/Universities
44. Search engines
45. Service information
46. Sounds
47. Stock quotes
48. Voice

People want better resources. A "List of Top Ten Oscilloscopes on the Market" is not enough to keep their attention. Powerful resources are being created to make it worth each person's while and to utilize the Web in all aspects of daily life. More powerful databases will keep the hobbyist/professional busy for hours. Secure online ordering will allow faster parts acquisition. Most of all, more people with similar interests will come together and create communities to help one another with their hobby/profession. Help from anywhere and anyone in the world — the TRUE "Information Age."

WHO IS PUTTING PAGES ON THE NET?

Just about everyone is contributing to the Net. Here is a small list:

1. Amateurs
2. Distributors
3. Engineers
4. Hobbyists
5. Manufacturers
6. Publishers
7. Retailers
8. Scientists
9. Students
10. Schools/Universities

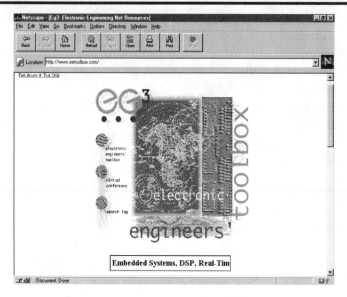

Figure 1. Many valuable electronics resources are available to you on the WWW. EG3 Communications has a top-notch site with plenty of useful information. http://www.eg3.com/

NOTE

University students with time on their hands and an account to burn are a great source for links and little bits of unknown data. They tend to gather information that seems useless to some but may be valuable to others.

WHAT YOU NEED TO ACCESS THE WEB

The vehicle used to tour the Web is not a surfboard; it is a program called a Web-browser. This device will drive you up and down the superhighway and tell you where the on and off ramps are. Taking a spin around cyberspace with a modern Web-browser will be a Cadillac ride compared to the Model-T browser versions of old.

The browser grabs the hypertext code from around the world and converts it into a useful format, which is displayed on your screen the way the designer wanted you to view it. The list of features in browsers is as big as the list of browsers available. Each person has preferences as to the software they use to surf the Web. Here is a brief list of popular Web-browsers and where to obtain the current freeware/shareware versions via FTP or WWW. Most Internet providers furnish you with a free browser in disk form. Load it up but be sure to shop around to see which browser best serves you.

PROGRAMS

Each of these programs has a Windows® 3.11/95/NT, OS/2, UNIX version for your needs. Lynx can be operated through DOS with a modem of any speed.

1. **Lynx**. A text-only based browser mostly used for computers with slow modems, slow processors, low memory, etc. People use it because text downloads much faster than graphics. Minimum system required to run efficiently: 286, MAC equivalent/ almost any speed modem.
   HTTP://www.w3.org/pub/WWW/Lynx/
   FTP://ftp2.cc.ukans.edu/pub/www/

2. **Microsoft Internet Explorer**. A great browser that is Netscape compatible and is currently freeware. It offers seamless operation with Windows Operating Systems. Minimum system required to run efficiently: 386-486, or MAC equivalent/ 8 megs RAM/ 14.4 modem.
   HTTP://www.microsoft.com/ie/ie.htm
   FTP://ftp.microsoft.com/msdownload/
   See site for current version.

3. **Netscape Navigator**. This is currently the most popular browser on the market, and rightly so. It's list of features and Netscape's experience have put it on the top of the heap. Netscape is currently shareware. Minimum system required to run efficiently: 386-486, or MAC equivalent/ 8 megs RAM/ 14.4 modem.
   HTTP://home.netscape.com/
   FTP://ftp1.netscape.com/

4. **Mosaic**. This original Web-browser, although lacking in features, is great for midrange PCs. Mosaic is currently freeware. Minimum system required to run efficiently: 386, or MAC equivalent/ 4 megs RAM/ 14.4 modem.
   HTTP://www.w3.org/pub/WWW/
   FTP://ftp.ncsa.uiuc.edu/pc/

Netscape Navigator and (to a lesser degree) Microsoft Internet Explorer were used to compile the addresses for this book. Most browsers would not have been able to correctly display 90% of the sites visited. Navigator and Explorer have the ability to enhance those sites to portray their full effects.

My personal preference is Netscape Navigator. It is a very professional program which perfectly fits my 486-100/12 meg/ 33.6 modem computer.

A small note about modems — don't worry about buying the fastest modem unless you will be doing huge file transfers. The servers around the world are loaded with traffic and a faster modem would not greatly improve your Web page load times. One hint is to use a provider which has what is called a T-1 or T-3 line. It means they can handle more traffic without being bogged down.

PREFERENCES

Setting up your browser's preferences can be difficult. Take your time. It is worth it. Learn and study the "options" list well. By utilizing each feature to the fullest you can save hours of search time. Here are a few common features to look for and learn.

1. **Open URL**. Lets you access HTTP, FTP and GOPHER sites.

2. **Bookmark**. Lets you place a theoretical mark on a page so you can come back to it later.

3. **Back/Forward/Home**. Lets you navigate sites and keep track of where you are.

4. **Email/Newsreader**. Applications built into Netscape and Explorer which let you send Email and read newsgroups without multiple programs.

Read each browser's help file and connect to their company's online help site for further details.

WEB TERMS YOU NEED TO KNOW

Homepage:
1. The first Web page on a Web site.
2. A Web page that describes an individual and their interests. Some companies use this term to create a "home-page-type feel" to their sites.

HTML:
HyperText Markup Language. The language used to create Web pages and link sites to one another.

HTTP:
HyperText Transfer Protocol. The protocol that lets the Web client communicate to the Web server.

HTTPS:
Secure HTTP. A server that is secure enough to send valuable information such as credit card numbers.
Example: HTTP://www.hwsams.com

Java:
A programming language that lets small executable programs and other items be placed within a Web page.

URL:
Universal Resource Locators. These are locations where documents reside along with the protocol needed to access them. A URL is like a street address to a company's area of the Internet.

Web Client:
Your Web browser.

Web Page:
An HTML document containing various links and objects.

Web Server:
The computer that the document resides on.

Web Site:
An area on the Web server which contains multiple Web pages.

WHAT IS SURFING?

"Surfing the Net" or "Surfing the Web" are the buzzwords for Madison Avenue. It means to slide around the sea of information and to have fun doing so. What better word to describe it? You go to one page on the Web, and while you are looking for a specific piece of information, you see a tantalizing morsel of data off to the side. "Oh! That looks interesting. I guess a quick jaunt to that page won't take long."

"Click!" goes the mouse.

You are now caught in the Web, surfing with no land in sight! You zoom over to the new electronics page. You see a note about how to make the best mixed drink this side of Mars. What this drink has to do with electronics is anyone's guess. Whoosh! The wave catches you and off you go until you find just what you were NOT looking for — a list of wrestlers over age 35. What happened to the datasheets you were originally trying to locate?

With hypertext documents, Web page designers can link any word or picture to any page on the entire Internet. This gives the ability to find what you are looking for as well as similar information, without having to go over every page on the Net.

Don't worry about a clock being around you when you surf; time ceases. The Universal Time Constant stops functioning at the moment you connect to your provider. Oh! A link to an Einstein page. Better check that out.

OTHER TIPS TO MAKE YOUR SURFING SMOOTHER

1. Use an 800 by 600 pixel resolution setting. Most pages are designed in this mode and won't look right in 640 by 480.

2. If your browser is loading a framed page and you can't see what is in the frame take the mouse and move the border over.

3. Learn the HOT-KEYS for your browser, then shut the toolbar, directory buttons and (if possible) the location window off. This gives you 20%-30% more viewing area with which to see Web pages.

4. Use your clipboard as a time saver, pencil saver, and sanity saver. Copy URLs from other documents to it, then paste them into your browser's Location Window.

WHAT CONSTITUTES A GOOD WEB SITE?

Each person has preferences as to what a good site is, but here are questions to help you make up your mind.

1. Is the site easy to navigate? Do you know where you are in the site at any given time?
2. Can you find the information you need fast?
3. Is there an easy-to-find Email address?
4. Does the page contain useful data among the regular advertisements?
5. Do you feel comfortable on the site? Does the site have a homey feeling?
6. Are there good visual cues to help you?
7. Are there links to other sites, or is the page a dead end?
8. Are there reasonable size graphics on the page? Huge graphics are modem and time eaters.
9. Are the graphics interesting to look at and pertinent to the page?
10. Are there complicated login procedures to get anywhere on the site?

11. How long have those *Under Construction* signs been up?

Sites can be a mind-numbing experience or a treasure to your eyes. Treat the great sites like gold and help the other sites by Emailing the creator (Webmaster) your comments, suggestions and opinions.

COMPANY SITES VERSUS HOMEPAGES

Internet worlds are cheap to make; all it takes is a deep pocket of creativity. Companies are not the only ones that can afford to publish web sites. Users can create homepages on a very limited budget. What is the difference between company sites and homepages?

Company Site: Generally an advertisement with links to company products. Good for finding a specific piece of information about that company's products.

Homepage: A personal statement of tastes and interests with links to these interests. These can be good if the creator of the page has similar interests to yours. They usually have done their Net homework and found all there is to know about a specific area of interest.

NOTE

Don't be fooled by the BIG COMPANY SITES. An old, reliable GM is more useful than a flashy BMW that is in the shop. Of course, it is nice to cruise the main drag in a fancy car once in a while.

WHAT IS CONTENT?

Content may be defined as *meaning* or *substance*. Does this site have useful information? Is it worth my time to really explore it? Does it have too many dead ends? If you can find pages that have content, the time you will save is exponential.

Don't worry if you found the information from some student's homepage from Keokuk or from Intel's Web site. Wading through useless garbage may become your new pastime; but, treasures await diehard data-diggers. Content is priceless.

WHERE TO FIND MORE INFORMATION ABOUT THE WWW

WWW FAQ@
HTTP://www.math.umass.edu/www_faq.html

SUMMARY

The World Wide Web will be your most used Internet resource while dealing with electronics and information searches. Learn the ins and outs well and save your sanity. With so many Web pages out there, you will need it.

The directory in this book contains mostly WWW addresses. Use these as a starting point and expand your knowledge.

TEACHING YOURSELF TO FISH

*"Give a man a fish and he will eat for a day.
Teach him to fish and he will eat for a lifetime."*
— Popular Quote

This book contains a directory to help you search for electronics information on the Internet. However, the most important chapter is this primer. It will teach you how to fish for data in this vast new sea of information. What resources are available to help you? How can you find the answers to very specific questions? Learn the skills of Net searches, and soon you will be pulling out boat loads of facts, figures, research and documents — an invaluable cache for your electronics hobby/profession.

HOW TO FIND WHAT YOU ARE LOOKING FOR IN THE SEA OF INFORMATION

"Up to 60 million documents on the Web? Hundreds of thousands of programs on FTP?? 500 Megabytes worth of Usenet articles a day??? How am I ever going to find what I am looking for?"

This information overload can be grounded with a good knowledge of search techniques. Learning the techniques *well* can save hours and also your sanity.

Search engines are in place to help you practice your newfound skills, giving you the ability to zip right to the piece of info that you want. This chapter will tell you what is available, and will teach you how to make use of the Net. Make sure to check out the TIPS section at the end of the chapter.

Most important of all — *Don't be discouraged*. If the Net doesn't provide you with exactly what you are looking for, it will most assuredly lead you to something that can.

HOW TO PERFORM AN INTERNET SEARCH

"I want to find a very specific fact from the Net. How do I do this?"

Internet searches are usually performed on a Web site called a *Search Engine*. A list of popular ones appears in this chapter. This site will present you with a box to type in *key-words* and additional qualifiers called *Boolean operators*. Hitting the search button after the text is entered will create a new web page with results. From there you are off and running to all points of the Web that relate to your query.

WHAT IS A SEARCH ENGINE?

Search engines are programs that search through many documents according to user-defined variables. They are accessed via a Web site. Place your query in the program, and out comes a list of possible matches.

Figure 1. DEC's Alta Vista is the most comprehensive search engine available to a web surfer. http://www.altavista.digital.com

WWW search engines allow the user to search Web pages that contain specific words or phrases. Using *keywords*, you can filter through every single page on the Web. Some engines can even look up every single word from the documents they catalog — most notably Open Text's search engine.

Other search engines, such as Yahoo, will allow a category search. You can pick from a list of topics of interest, follow the chain of subcategories, and *bingo*! You have narrowed your page choice down to only a few.

Each search method has its pros and cons. One engine may be just right for speeding you along, yet slow and tedious for other search purposes. See the EXAMPLES section for further help.

WHAT IS SURFING AND SHOULD YOU BE DOING IT?

Surfing is a media buzz word. It means to flounder around the vast amounts of decentralized data in the Internet — an aimless search. For example, go to a page titled, "How to program a PIC16C84." On the same page there is a link to NASA space photos. Click! Off you go to page after unrelated page until you can't remember where you began. It's great for beginners, but get it out of your system early. It is unproductive.

WHAT TO DO ONCE YOUR SURFING PHASE IS OVER

The surfing you did as a beginner did serve a purpose. Hopefully, you had a pen and a pad of paper in hand to jot down little treats. Maybe you kept bookmarks of a few interesting sites in your browser. Take those notes and bookmarks and look for keywords. Those keywords can now be plugged into a search engine, and some serious data digging may now occur.

SEARCH TERMS THAT YOU NEED TO KNOW

1. **Boolean**. The logic that search engines and computers are based upon.

2. **Boolean Operators**. AND, OR, NOT, NEAR. Use between keywords to narrow searches.

3. **Category Search**. A search that uses indices to refine hits.

4. **Decreasing Order of Confidence**. See *Ranking*.

5. **Engine** or **Search Engine**. The program or algorithm that sifts through many documents. It then produces results that guide you to what you are looking for.

6. **Field**. An area you type text into.

7. **Hits**. Pages that meet the criteria that you plugged into the search field.

8. **Indices**. Plural for *index*. A type of search engine that lists categories, then branches to subcategories.

9. **Queries**. Internet searches. A combination of your keywords and the operators.

10. **Ranking**. To put a list of hits in the order that best suites your search criteria; the highest being the most likely match. Some search engines use a scoring system with 100% being the site that most meets what you were looking for. Each engine is different.

11. **Scoring**. See *Ranking*.

12. **Search Field**. The area in which you type keywords and operators.

13. **Wild Card**. A symbol that stands for one or more characters. Used to let you look up words that contain a certain unknown element. Example: electr* would access files with electronics, electrical, etc.

BOOLEAN SEARCH ENGINES

George Boole gives you a private tour of the Internet using his mid-19th century discovery, *Boolean Logic*. The same logic that runs your PC speeds you along the Internet and cuts down your search time to minuscule jaunts.

By using logic operators, a Boolean search can filter searches down to a few specific pages. Common operators are AND, NOT, OR, NEAR.

Each Boolean-based search engine uses different algorithms to help with completing your searches. When you go to a Boolean-based page, you are presented with a space in which to insert text. This is where you type keywords and Boolean operators. Once the field is complete, hit a search button. A new page full of addresses will be returned, which contain your keyword. A starting point.

Basic Boolean Operators

AND. Can also use a *space* or *+*.
Allows the search of multiple words in one document.

Examples:
resistors AND color AND code
Will return documents containing those three words.

Or ~ *resistors color code* (note: the AND's are added automatically by most engines)

Or ~ *resistors + color + code*

NOT. Can also use *NO* or -
Excludes the next word from a document.

Examples:
multimeter NOT dmm
This would exclude DMM's from your search results.

Or ~   multimeter NO dmm

Or ~   +multimeter - dmm

OR
Will include documents with one word or another.

Example:
digital OR analog
Will return pages with the keywords *digital* or *analog*.

NEAR
Will return a list of documents that have two words close to each other. Usually within 10 words.

Example:
schematic NEAR stereo.
This would return a document with the sentence, "This schematic of Pioneer's new stereo is now available via FTP."

Boolean Plus Points
1. You can narrow a search down to only a few hits using Boolean.
2. These engines are able to search most of the WWW's 60 million odd pages at once.

Boolean Bad Points
The amount of data you are searching through is usually more vast then indices — thus harder to narrow and sometimes less accurate.

Boolean Uses
Good for trying to locate exacting or hard to find data.

NOTE

Use small letters in all keywords, unless you are abso-
lutely sure which letters are capitalized. Most engines
will automatically search for upper and lower cases if
you use all lower case, but will not take an upper case
and make it lower case.

Example

Resistor will only return documents with the word *Resistor* in them. Using *resistor* will return documents with
resistor and *Resistor* in them.

INDICES

Indexed sites are simple and fun to use. For example, when
you go to Yahoo's site, you are presented with a list of categories and a search field. You can now search in one of two
ways:

A. Pick a category and follow it to a list of subcategories.
Branch down until you reach the URL needed.

Example:
Follow this path page after page until you arrive at given
the URL:
Start with *SCIENCE* to *ENGINEERING*, to *ELECTRI-
CAL ENGINEERING*, to *CIRCUITS*, to *INTEGRATED
CIRCUITS — DIGITAL,* to http://www.mrc.uidaho.edu/
vlsi/vlsi.html

OR

B. Type a keyword into a search field, and the engine will
return the categories (or subcategories) at which desired
pages exist, as well as other URLs in the area.

Example:

Type *vlsi* in the search field upon entering Yahoo's main site. Out will come a listing of categories and sites containing the keyword *vlsi*. In the list will be http://www.mrc.uidaho.edu/vlsi/vlsi.html

Plus Points of Indices

Categories make it easier to browse similar items.

Bad Points of Indices

There are a limited number of pages cataloged.

Uses for Indices

To search for information quickly in a given topic or category.

HOW TO OPTIMIZE YOUR SEARCHES

There are a few more all-purpose skills needed narrow your searches. Remember, the more keywords and operators you use, the greater the accuracy of finding exactly what you are looking for.

PHRASES. By placing a phrase in quotes, you will be able to search for a list of documents containing the exact combination of words.

Example:

Typing into the search field, "How do you etch a printed circuit board," may return a FAQ about board etching.

When you type the sentence with or without quotation marks, the search engine will see each word with AND operators in between words without the quotes. I.E.: how AND do AND you AND etc. It would then return documents with those words scattered throughout — not in series.

WILD CARDS. By using a wild card (*), you can search for items that contain a snippet of information.

Example:
> *LM4** may return all the LM4000 series part numbers in a company database. The * is the wild card.

This feature can also be used if you are not quite sure how to spell a word.

Example:
> You want to find a product called a *cyberemulator*, but you only remember the *emulator* part. **emulator* will return all names with *emulator* as the suffix, such as *cyberemulator.*

Each engine has different symbols to represent wild cards.

SIMPLE and **ADVANCED QUERIES**. Sites such as Alta Vista offer the option of a simple or advanced query. Make use of the method that will save you the most time; i.e., don't use a complex search just to find common data, and don't use simple searches for something that is inherently complex to look for.

PARENTHESES. When using operators, it may be best to use parentheses to differentiate a statement, just as you would in algebra. Example: NOT (ram AND "72 pin") to exclude *72 Pin RAM* from your search.

MORE SEARCH TRICKS

1. By typing the name of a company into the URL field in Netscape Navigator or Microsoft Internet Explore, you may be able to access the company's URL.

Example:
> Type in "Intel".
> Netscape will load up URL http://www.intel.com/

2. Check magazines and commercials for URLs.

3. Homepages made by individuals and not companies, schools or organizations, will have more links to perpetuate your search. Seek them out when someone else is interested in your search.

4. Ask friends for their bookmark files.

5. Some sites will have their own site search engines. Utilize them whenever possible.

6. Sometimes you will find a domain that contains numerous electronics links. Some domains actually specialize in links, which can be very useful.

7. When doing searches, it will help with Boolean searches if you know the company name or source. Just plug it in the search form.

Example:
   "Electronic stores" AND "Toronto" AND "Active" AND "Victoria pk."

8. By taking directories off the URL, you will be able to view similar data.

Example:
   Removing *digital* from http://pobox.com/~electronics/digital/ will allow you to access a whole site devoted to electronics.

9. Write down or bookmark the addresses you have stumbled upon. Some of the most valuable sites you come across may originate from stumbling.

10. Surfing is great, but not as a starting point in finding data on the Net. A better scenario would be to start with a search engine, then surf the sites suggested.

11. Find the HELP file on each search engine you use. Learn the commands, tricks and tips that each one uses.

WHAT, BESIDES SEARCH ENGINES, CAN YOU USE TO PERPETUATE YOUR SEARCHING?

Sometimes a "John Smith's Projects Page" will offer more content per pound than a huge corporate site. Really go through these types of sites as they may contain useful information that can help you with your searches.

WORLD WIDE WEB SEARCHES

The Web is IT! Celebrity status has been reached overnight. Almost every type of business, resource, school and hobbyist is cranking out Web page after Web page, trying to become part of the fanfare. With thousands of Web sites being added daily, it has become next to impossible to find even the simplest fact. Enter Web search engines, the true stars of the WWW.

HOW MANY WEB PAGES ARE ON THE WORLD WIDE WEB?

Unknown. Current cataloged page numbers are listed below to help give you an idea.

NOTE
These figures were current at the time of this book's writing and are the claims of each individual search engine.

---

1. Alta Vista claims 30,000,000 in their database.
2. Excite claims 50,000,000 in their database.
3. Lycos claims 51,000,000 in their database.
4. HOTBOT tops out, claiming 54 million individual URLs, with an estimated 6 million more that are not cataloged.

CAN YOU SEARCH ANY OTHER RESOURCE THROUGH THE WWW?

Most definitely! Using Web-based search engines, you can search for:

1. Web pages.
2. Newsgroup articles.
3. Gopher.
4. FTP.
5. Email addresses.
6. Company and personal addresses, phone numbers, etc.
7. Maps.
8. Libraries.
9. Dictionaries.
10. Directories of all sorts.
11. Pictures, movies, sounds.
12. Government databases.
13. School databases.

COMMON WWW SEARCH ENGINES

This is a brief list of search engines available to you. Each of these sites has its good and bad points. I will leave the choice up to you.

100 Hot Websites
If you are looking for the top 20 computer sites, or just want to enjoy what the Web has to offer, check out this engine:

HTTP://www.100hot.com/

Alta Vista
The main advantage of this engine is its lack of advertisements — which equals speed. This is possibly the most popular engine because of the lack of ads. If you want to find something and know how to use Boolean, this site will find it. You can search WWW or Usenet:

HTTP://www.altavista.digital.com/

Awebs Newsgroup Archives
Searches last month's newsgroups:

HTTP://awebs.com/news_archive

Bigbook
Yellow pages directory/search engine used to locate businesses. Includes maps to locations:

HTTP://www.bigbook.com/

Bigfoot
Use to search for someone's Email address, phone number or street address. Includes a zoomable map:

HTTP://www.bigfoot.com/

DejaNews
Searches Usenet articles back to March, 1995. Uses up to 80 Gbytes (50 million articles) of searchable data:

HTTP://www.dejanews.com/

Download.com — By CINET
Downloads all of the latest and most popular files:

HTTP://www.download.com/

The Electronic Library
Searches through various newspapers, magazines, newswires, books, maps, photos and works of literature contained on the Net. You can also pose questions for the search engine to answer. A great, quick research tool:

HTTP://www.elibrary.com/

Excite
Uses a standard search algorithm and a concept engine to give you hits of similar categories. The "more like this" feature is great. Can search WWW, Web site reviews, Usenet,

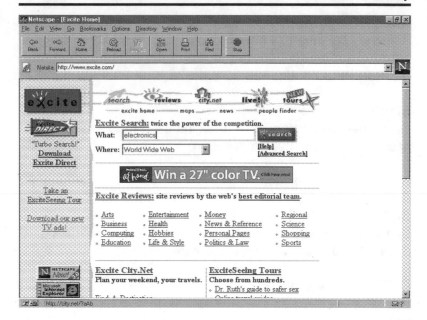

Figure 2. Find a search engine that you are comfortable with. Excite is packed with useful search tools for the electronics hobbyist or professional.

Usenet classifieds, maps, news, tours, references, Email, dictionaries, maps, and more. This is a great, all-around search engine:

HTTP://www.excite.com/

Four11
Searches for Email addresses, telephone numbers, and Netphone numbers for persons, business, government, celebrities:

HTTP://www.four11.com/

GTE SuperPages
Searches maps, classifieds, Web sites, Yellow Pages with maps, and business Web directories:

HTTP://www.superpages.com/

HotBot
Has Access to 54 million documents, and searches WWW and Usenet. Very well thought-out search rankings:

HTTP://www.hotbot.com/

Infoseek
Searches for directories and similar listings in WWW, Usenet, FTP, Gopher, Email addresses, News, and FAQs:

HTTP://guide.infoseek.com/

Lycos
Has access to 51 million unique URLs. The ultimate multimedia site. *Very* popular. Searches the WWW for sounds, pictures and video, and can also search by subject. Also has People Find and road maps, and lists the top 5% of the most popular sites:

HTTP://www.lycos.com/

Magellan
Browses topics or searches. Includes WWW, FTP, Usenet, Gopher, Telnet. Also has People Finder, Search Voyeur, Net Events, and Stellar Sites. This engine is small but offers site reviews to help save you time:

HTTP://www.mckinley.com/

ON'VILLAGE
Has access to 15 million businesses and organizations, including phone numbers, addresses, and maps:

HTTP://www.onvillage.com/

Open Text
Searches through every word listed in a document. Very useful when you are looking for specialized data:

HTTP://www.opentext.net
HTTP://index.opentext.net

Reference.com
A Usenet and mail list archive, with 16,000 newsgroups cataloged:

HTTP://www.reference.com/

SEARCH.COM — By CINET
Contains multiple search engines and other little treats:

HTTP://www.search.com/

SHAREWARE.COM — By CINET
The place to find freeware and shareware:

HTTP://www.shareware.com/

The Usenet Newsstand
Searches through various newsgroups:

HTTP://criticalmass.com/concord

Yahoo!
The pioneering Web engine. A guide controlled by using categories. It also has limited Boolean searches. Great entertainment links, plus Yellow Pages, person searches, maps, stock quotes, and sports scores:

HTTP://www.yahoo.com/

WebCrawler
A small search engine that is Boolean-based. Very simple to operate:

HTTP://www.webcrawler.com/

My personal preference is Alta Vista because of the shear speed of Digital's search computers. It is very versatile. When searching for "common" information, I prefer Yahoo or WebCrawler. When I have only a few extra moments to sit and enjoy the Web, I use Lycos or Excite. See the section

EXAMPLES to get a better idea which to use for a specific purpose.

FTP AND GOPHER SEARCHES

ARCHIE for FTP

By using Archie, you can search millions of files available via FTP. However, Archie searches only for file names. WWW-type search variables can be used, plus you can input other information to narrow a search; i.e., the date in which the file was created, a word in the path, the size of the file, etc. By using wild cards, you can find the exact file you are trying to locate.

If you know the exact name of the file you need, or even a section of the file name, this is the quickest resource.

Two good Archie programs are *fpArchie* and *WS-Archie*. Both available from Tucows at HTTP://www.tucows.com/.

HINT
Narrow your searches by inputting your preferred O/S into the path field. *Example*: WIN95, OS/2, MAC, etc.

GOPHER SEARCHES

Using WS-Gopher you can access information servers for electronics, libraries, government publications, phone books you name it. On the Gopher servers you will be able to use keyword searches in different categories.

Example:

Access gopher.uiuc.edu and find *Computer, Documentation* and *Software.* You will see *Keyword Search of Internet Documentation.* From there, you will be able to do a search for electronics documents, etc.

WS-Gopher is available on Tucows at
HTTP://www.tucows.com/

You can also access Gopher servers using your Web
browser. Some WWW-based search engines will even
search Gopher for you.

INTERNET FISHING TIPS

1. Use the clipboard as a time saver, pencil saver, sanity
    saver. Take URLs from browsers, IRC, or anywhere.
    Paste them into the browser's URL field.

2. If it is not listed in this book do a WWW search or check
    out http://pobox.com/~electronics/ for tips.

3. Use lower case letters in searches unless you are abso-
    lutely certain of the placement of any capital letters.

4. Want to know more about a company? Internic at HTTP://
    www.internic.net/ has all of the contact information, ad-
    dresses, and phone numbers that you will probably need.
    Search for their URL with HTTP://rs.internic.net/cgi-bin/
    whois

5. Find part numbers by plugging the number and
    manufacturer's name into Alta Vista or site search en-
    gines.

6. Try to keep a sense of where you are inside a site in order
    to navigate more easily.

7. Look at distributor linecards for links.

8. Don't be afraid of Emailing companies for help in your
    searches. Sales and tech support people are seldom
    offended by your questions.

9. To find a word like *Mississippi*, use a wild card.

Example: mis*pi

10. Use capitals to find a company, if you know the exact placement.

Example: "Acme Electronics."

11. If a search is coming up with too many resultant pages that have nothing to do with what you want, add NOTs to the field.

Example:
"Electronics stores" NOT "Consumer electronics" would exclude electronic stores that have nothing to do with your hobby.

12. Use www.shareware.com when searching for programs.

13. mIRC has a feature called a *URL Catcher* that will "grab" URLs that people type into channels. You can open up your browser to a page right from mIRC without typing a thing. Skim through it once in a while to find interesting sites.

14. Don't go cold turkey on surfing. Write down sites that you may find useful or interesting in the future, and use your list as a reference while doing actual searches.

15. Think of words that someone may use to describe the product you are searching for and plug them into the search field. Remove incorrect words until you find what you need.

THE FUTURE

A search engine learns from *you*. The more people who use it, the more advanced the algorithms become, which can help you in the future; the search engine develops while you are developing your own search skills. One day the engine will be smart enough to predict what "Joe Smith" is likely to be looking for, and go right to it. This is not yet true of search engines, but just look at how far the Internet has already come in only three years.

EXAMPLES

Here are a few example searches to help you with your own electronics searches:

A. You are searching for a products datasheet, the PIC16C84, and you have no idea where to start. You try typing *pic* into Alta Vista. Out come 1,226556 hits. There goes the next few weeks. Next, you type in *pic16c84*. Only 579 hits. Still way too many! You then decide to go to Advanced Queries and into the Selection Criteria Field — "pic16c84 datasheet". Bingo! Only three possibilities, and two of them are exactly what you want.

B. You are looking for general information on PCB design On Yahoo, you go from *Science* to *Engineering*, to *Electrical Engineering*, to *PC board Design*, to a list of a few sites with which to begin. You visit a few of these sites and in them are a few links to other PCB resources.

C. You are looking for a picture of the newest Pentium chip. Onto Lycos' picture searching engine you go. After typing in *Pentium Chip*, you are presented with a few sites in which you can view a picture of the processor.

D. You have lost your friend's Email address and need to ask him for some electronic schematics. You go to Excite, then to Email Lookup, and type in his name and the city he lives in. Within seconds, you have his new Email address.

E. You are fixing an RCA VCR and are encountering problems with the power supply. You begin to wonder if anyone else has had this problem. You type *rca AND vcr* into the Quick Search field and find a list of articles that contain those two elements. On it you find an article from someone who had the same problem, but there are no responses. Go back to the Quick Search field and type in the exact article name.

This will bring back the original article and the responses of people who helped him, which can help *you* fix your VCR.

THE WEB DIRECTORY: AN INTRODUCTION

The World Wide Web's explosive growth in recent years has brought millions of companies, organizations, schools and individuals flocking to this new media tool. As a result, the effects on the Internet have been dramatic as older applications are constantly being replaced by newer ones. As an ever-changing and renewable source of information, the Web has become an all-purpose application.

This directory focuses mainly on electronics sites on the WWW. Please note that the Web is constantly changing, as new applications replace older ones; so the sites in this listing may change. Included in the electronics listings are:

1. The company or resource name.
2. A brief description of the company or resource.
3. The information currently found on that site.
4. A contact address.
5. The WWW URL to access that web site or page.
6. An anonymous FTP URL to access files — if applicable.
7. A 1-to-10 rating, 10 being best.

My own site has a search tool to help you find information on a particular category. For example, if you are looking for all sites containing PIC data, the tool will give you a listing of them, by name only. Then you simply consult this book to find the URLS. My *Electronics Guide Site* is at:

HTTP://pobox.com/~electronics/

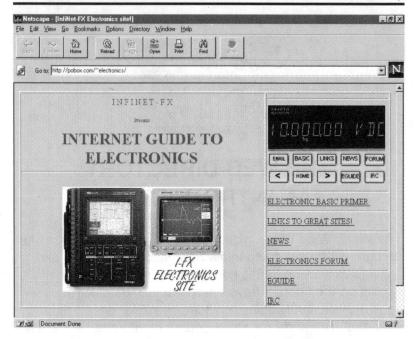

Figure 1. The author's electronics-related Website. With basic and advanced help for hobbyists — links, IRC information, etc. See it at http://pobox.com/ ~electronics.

Great care was taken in creating the directory for this book, but if a URL is incorrect or cannot be found, take a peek at the site listing for corrections. Please keep in mind that the Internet is an ever-changing entity, and it is somewhat hard to keep up with it. Learning search techniques will make you invincible to the constant changes. If you know of any other electronics-related Internet sites or resources that need to be added to future editions of this book, please submit them to me at electronics@pobox.com.

NOTES

1. A 10-plus rating is a top-notch Web site.
2. The INFO line is a sample of what is available at that site.
3. Typing HTTP:// is unnecessary with most browsers.
4. UNIX is case sensitive on its directory structure. If a site says —
   HTTP://www.site.com/HTML/index.html
   — then make sure the HTML is capitalized.

5. UNIX sees a difference between .htm and .html — make sure to use the correct one.
6. If a site's address has changed do a search using their name.
7. By contacting a Web Master, you will be able to get employee Email addresses.

DISCLAIMER

The directory contains a rating system to help with your surfing choices. Site ratings are based on the overall impression that I gleaned after 5 to 10 minutes of viewing. Here are the factors listed in decreasing order of importance. Please note that the content will always outweigh every other factor.

1. **Content**. Was the page worth my time? Did I come out of it learning something? If it was a company site, did it explain the products to my satisfaction? Did the amount or quality of resources on it *amaze* me?
2. **Ease of navigation** and ease of locating information. Did I know where I was within the site at any given time? Did I find what I was looking for easily?

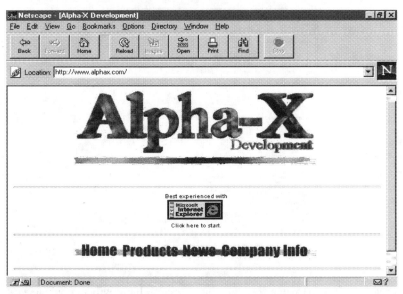

Figure 2. Simple company Web sites, such as Alpha-X Development's, are a pleasure to navigate. http://www.alh.

3. **Links**. Were there any links to similar information?
4. **Inventiveness**. Was the page new and exciting, or a carbon of one seen as little as 30 minutes ago?
5. **Loading Time**. Did the page load fast enough? Were bandwidth and other technical factors taken into consideration?
6. **Graphics**. Did it have interesting graphics to keep my attention, and were the graphics pertinent to the page?

SITES WERE <u>NOT</u> RATED ON:

1. The financial statements of the company or resource. It didn't matter to me if the site was a multimillion dollar company or a student's home page.
2. The size of the site.
3. The amount of advertisements in the site. Advertising supports the Net's backbone, and they are fine if "moderated." However, I did evaluate advertisements based on whether or not they told more than a magazine or TV ad.
4. The artistic quality of the graphics. More attention was given to the fact of whether or not the graphics were pertinent to the site or too large.

Figure 3. It won't win a "best graphics" contest, but it will sure win a "best content" contest. Filip Gieszczykiewicz's site proves that you don't need top-rate graphics to have a great site. http://www.paranoia.com/~filipg/.

251 FAQ
Intel 251 uC info from Keil
EMAIL: support@keil.com
HTTP://www.keil.com/251faq.htm
RATING: (7)

2600 Magazine
The Hacker's Quarterly
INFO: Covers, pay phones, news,
you never know what will show up!
EMAIL: webmaster@2600.com
HTTP://www.2600.com/
RATING: (9)

3Com
Network products manufacturer
INFO: Product info/support/
network solutions
EMAIL: See site for locations
HTTP://www.3com.com/
RATING: (9)

3Dlabs, Ltd.
info@3dlabs.com

3DO Systems & Studio 3DO
Video game systems and software
INFO: Product info/news/previews
EMAIL: See site
HTTP://www.3do.com/
RATING: (9)

3E Technology, Inc.
Search/integration of electronic
parts/components
INFO: Intro, categories, vendors list
EMAIL: three@ios.com
HTTP://www.3etech.com/
RATING: (5)

3M
Materials innovator — fiber optics,
medical instruments, and 60,000
more products

INFO: Market centers/products/list
of products
EMAIL: innovation@mmm.com
HTTP://www.mmm.com/
RATING: (9)

4QD
Manufactures speed controllers
INFO: FAQ on speed controllers,
products, and tons of electronics
related content! Well worth your
surf time.
EMAIL: 4QD@argonet.co.uk
HTTP://box.argonet.co.uk/users/
4qd/
RATING: (10)

68HC11 FAQ
Bob Boys' Motorola Chip FAQ
SEE: FAQ Library — Electronics

8031 PCB Page
Maker of PCB boards for 8031
INFO: Links, PCB info
EMAIL: jbrown@whynet.com
HTTP://users.why.net/jbrown/
pcb.htm
RATING: (8)

8051 FAQ
SEE: *FAQ Library* or *Russ Hersch's
8051 FAQ*

8X8, Inc.
Video compression ICs/systems
INFO: Products, solutions, links,
news
EMAIL: webmaster@8x8.com
HTTP://www.8x8.com/
RATING: (8)

A

A C Trading
Zurich based distributor of parts
INFO: Online catalog and database
EMAIL: pk@ac-electronics.ch
HTTP://www.ac-trading.com/
RATING: (7)

A.M.S. AlltronICs, Inc.
Ontario-based supplier
INFO: Company info
EMAIL: ams@icacomp.com
HTTP://www.ica.net/pages/ams/
RATING: (6)

AAEON Electronics, Inc.
Single board computers, PC/104
modules, ODM/OEM
INFO: Product info/support/specs,
etc.
EMAIL: info@aaeon.com
HTTP://www.aaeon.com/
RATING: (8)

Aavid Thermal Technologies, Inc.
Electronic cooling methods
(heatsinks)
INFO: Reps, product info/specs,
news
EMAIL: forreste@aavid.com
HTTP://www.aavid.com/
RATING: (8)

Abbott Electronics
Industrial & military power
supplies
INFO: Company site, employment,
etc.
EMAIL:
sales@abbottelectronics.com
HTTP://abbottelectronics.com/
RATING: (7)

Abletronics
California-based supplier
INFO: Company site
EMAIL: ablesales@abletronics.com
HTTP://www1.abletronics.com/
abletronics/
RATING: (6)

Absolute Value Systems
Wireless video teleconferencing
INFO: Slow scan TV/ham radio
info, links, ham
EMAIL: johnl@world.std.com
HTTP://www.ultranet.com/~sstv
RATING: (9)

Abstract Technologies, Inc.
IBM RISC System/6000 Peripherals
INFO: Product, support, info on
IBM, etc.
EMAIL: info@abstract.com
HTTP://www.abstract.co.nz/
RATING: (7)

Accent Electronic Concepts Intl.
Components
INFO: Company site, listing of
components
EMAIL:
102121.1676@compuserve.com
HTTP://www.well.com/user/
business/aec.html
RATING: (6)

Accolade Design Automation, Inc.
Electronic design automation
software
INFO: Product info, free CD,
VHDL made easy, links
EMAIL: sales@acc-eda.com
HTTP://www.acc-eda.com/
RATING: (10)

Accton
Computer networking products
INFO: Ethernet/token ring product info/drivers/links
EMAIL: support@accton.com
HTTP://www.accton.com/
RATING: (9)

Accurate Automation Corp.
Artificial intelligence and neural network design
INFO: Products, projects, jobs, links
EMAIL: marketing@accurate-automation.com
HTTP://www.accurate-automation.com/
RATING: (9)

Accurite Technologies, Inc.
F. Drive Diagnostics and PCMCIA products
INFO: Tech papers for floppy service/repairs
EMAIL: info@accurite.com
HTTP://www.accurite.com/
RATING: (9)

Accutronics International
Board-level components supplier
INFO: Company site
EMAIL: tcarmody@earthlink.net
HTTP://www.accutronics.com/
RATING: (4)

Acer, Inc.
Computer manufacturer
INFO: Regional/product info, contests, news
EMAIL: ai_webmaster@acer.com
HTTP://www.acer.com/
RATING: (9)

Acme Electric
Power conversion products
INFO: Company page
EMAIL: acme@moran.com
HTTP://www.moran.com/htmld/acme.html
RATING: (6)

ACNI Electronics, Inc.
Part tracking of unavailable components
INFO: General company info
EMAIL: acni@snj.com
HTTP://www/snj.com/acni
Also

A1 Memory Now!
Discount memory outlet
INFO: Stock info and credit card retrieval
EMAIL: memory@snj.com
HTTP://www.snj.com/memory
OVERALL RATING: (8)

Acopian Technical Company
Power supply manufacturer
INFO: Product selector tool, specs/photos
EMAIL: n/a
HTTP://www.acopian.com/
RATING: (7)

Actel Corp.
FPGAs, PLDs, etc.
INFO: Products apps/basics. Huge company site!
EMAIL: webmaster@actel.com
HTTP://www.actel.com/
RATING: (10)

Action Electronics
Distributor of electronic parts/equipment

A

INFO: Quick answer email, Linecard, links
EMAIL: action@edm.net
HTTP://www.action-electronics.com/
RATING: (7)

Adaptec, Inc.
Adaptec-designed hardware & software products are used wherever data must be transferred from a computer to a peripheral device or network.
INFO: Company, product, promotion, support and employment
EMAIL: webmaster@adaptec.com
HTTP://www.adaptec.com/
FTP://ftp.adaptec.com/pub/
RATING: (10)

ADC Telecommunications
Telecommunications supplies
INFO: Company site
EMAIL: technical@adc.com
HTTP://www.ps-mpls.com/adc/
RATING: (7)

Adcom
Sound equipment
INFO: Product info, contacts
EMAIL: adcom@soundsite.com
HTTP://www.soundsite.com/adcom/
RATING: (8)

Addison Wesley Longman, Inc.
Learning resources — Huge site
INFO: Education info, division info
EMAIL: webmaster@awl.com
HTTP://www.aw.com/
RATING: (10)

Ademco and Divisions
Security products
INFO: Product info, support
EMAIL: webmaster@ademco.com
HTTP://www.ademco.com/
RATING: (9)

Adtran, Inc.
Network systems
INFO: Product info/papers, ISDN, T1.
EMAIL: tpage@adtran.com
HTTP://www.adtran.com/
RATING: (9)

Advacom
Industrial electronics distributor
INFO: Company info, links, etc.
EMAIL: info@advacom.com
HTTP://www.advacom.com/
RATING: (8)

Advance International Group, Ltd.
Manufactures power supplies/components/test instruments
INFO: Product info/specs/pics
EMAIL: 106056.604@compuserve.com
HTTP://www.dircon.co.uk/advance-int/
RATING: (7)

Advanced Hardware Architectures, Inc. (AHA)
IC manufacturer
INFO: Company site
EMAIL: sales@aha.com
HTTP://www.aha.com/
RATING: (7)

Advanced Logic Research, Inc. (ALR)
Computer manufacturer

INFO: Product support, links, you name it!
EMAIL: See site for directory
HTTP://www.alr.com/
RATING: (10)

Advanced Micro Devices, Inc. (AMD)
Microchip manufacturer
INFO: Extensive company site with product sheets, apps, datasheets.
EMAIL: See site for directory
HTTP://www.amd.com
RATING: (9)

Advanced Micro Systems, Inc. (AMS)
Tools for automation
INFO: Software, apps, stories.
EMAIL: lpf2000@aol.com
HTTP://ams2000.com/
RATING: (8)

Advanced RISC Machines (ARM)
HTTP://www.arm.com/
Unable to access at time of reviewing

Advanced System Products, Inc.
SCSI adapter manufacturer
INFO: Product info/papers/software, contact
EMAIL: sales, support@advansys.com
HTTP://www.advansys.com/
FTP://ftp.advansys.com/pub/
RATING: (7)

Advanced Technology Marketing, Inc. (ATM)
Independent stocking distributor
INFO: Company site

EMAIL: info@advanced-tech.com
HTTP://www.advanced-tech.com/
RATING: (7)

Advanced Transdata Corp.
PIC16/17 development tools
INFO: In-circuit emulators and programmers for PIC16/17 MCUs, EPROM and other MCU programmers
EMAIL: atc1@ix.netcom.com
HTTP://www.adv-transdata.com/
FTP:// ftp.dfw.net/pub/users/atc
RATING: (8)

Advent Electronics
Parts distributor
INFO: Company site with some links
EMAIL: n/a
HTTP://www.adventelec.com/
RATING: (6)

AEG Schneider Automation
PLCs & motion control systems
INFO: Support, online docs
EMAIL: See site for forms
HTTP://www.modicon.com/
RATING: (8)

Aegis Electronic Group
Distributor of video products
INFO: Product info and request forms
EMAIL: aegis@primenet.com
HTTP://www.aegis-elec.com/
RATING: (7)

AEI Electronic Parts
Parts distributor
INFO: Company info
EMAIL: n/a

A

HTTP://members.gnn.com/hca/aei/
RATING: (6)

AESCO Electronics
Electronics parts distributor
INFO: Company site
EMAIL:
webmaster@www.aesco.com
HTTP://www.aesco.com/
RATING: (5)

Agastat
Interconnection components
HTTP://www.industry.net/agastat/

Agile Networks
Multiprotocol intelligent switches
INFO: Product info/support/docs
EMAIL: info@agile.com
HTTP://www.agile.com/
RATING: (7)

AGV Products, Inc.
Automated guided vehicle systems/
components
INFO: Online catalog with specs,
pics
EMAIL: agvp@vnet.net
HTTP://www.vnet.net/agvp/
RATING: (6)

AIM Technology, A Network General Company
Performance benchmarks for UNIX
and Win-NT machines
INFO: Products, support, certified
reports and performance results on
hundreds of current systems
EMAIL: benchinfo@aim.com
HTTP://www.aim.com
RATING: (8)

AirBorn Electronics — Australia
Designer of microprocessor
electronics
INFO: Everything for uC and uP
designing! Great
information/resource site
EMAIL: stevenm@airborn.com.au
stevenm@zeta.org.au
HTTP://www.airborn.com.au/
HTTP://www.webcom.com/airborn/
RATING: (10)*****

AITech International, Inc.
Voice Recognition and other
telephony technology
INFO: Products, apps, demos
EMAIL: info@aitech.com
HTTP://www.altech.com/
RATING: (9)

AJAX Electric Motor Corp.
Manufactures electric motors
INFO: Product info, distributor
EMAIL: sales@ajaxmotors.com
HTTP://www.ajaxmotors.com/
RATING: (7)

Akugel Electronics Development
Hardware development and advice
INFO: Developer's forum, company
links.
EMAIL: akugel@t-online.de
HTTP://www.webcentre.com/
akugel/
RATING: (7)

Alberta Printed Circuits Ltd.
Prototype PCB fabricators
INFO: Tons of links, tips, programs
EMAIL: staff@apcircuits.com
HTTP://www.apcircuits.com/
RATING: (8)

Alden Electronics, Inc.
Weather system electronic products
INFO: Product info, great links to weather stations
EMAIL: info@alden.com
HTTP://www.alden.com/
RATING: (9)

All-in-One Search Page
Synthesized search tool
INFO: Extensive search tool of interest to hobbyists/professionals
EMAIL: wcross@albany.net
HTTP://www.albany.net/allinone/
RATING: (9)

All Electronics Corp.
Electronics parts house
INFO: PDF files of complete catalog or sections, buying info, product pics
EMAIL: allcorp@allcorp.com
HTTP://www.allcorp.com/
RATING: (8)

Allen-Bradley
Rockwell Automation
INFO: Product info, data, contacts, etc.
EMAIL: See site for directory
HTTP://www.ab.com/
RATING: (10)

Alliance Electronics, Inc.
INFO: Company site, a few links
EMAIL: n/a
HTTP://www.allianceelec.com/
RATING: (3)

Alliance Manufacturing Software
Windows-based manufacturing software
INFO: Product info/support, forum
EMAIL: sales@alliancemfg.com
HTTP://www.alliancemfg.com/
RATING: (7)

Alliance Semiconductors
Semiconductor manufacturer
INFO: Company info, data sheets
EMAIL: webmaster@alsc.com
HTTP://www.alsc.com/
RATING: (7)

Allied Electronics, Inc.
Electronic components & equip. distributor
INFO: Online stock and order
EMAIL: webmaster@allied.avnet.com
HTTP://www.allied.avnet.com/
RATING: (10)

AlliedSignal, Inc.
Aerospace, automotive, fibers, plastics, etc.
INFO: Each division's products
EMAIL: See site for listing
HTTP//www.allied.com/
RATING: (8)

Allison Technology Corp.
Digital storage oscilloscope manufacturer/distributor
INFO: Apps, prices, briefs
EMAIL: atc@accesscomm.net
HTTP://www.atcweb.com/
RATING: (8)

Alltronics
Buys and sells electronics and equipment
INFO: Specials, hobby corner, ordering info

A

EMAIL: ejohnson@alltronics.com
HTTP://www.alltronics.com/
RATING: (9)

Almo Wire & Cable
Distributor
INFO: Company site
EMAIL: Email form on site
HTTP://www.almo.com/
RATING: (7)

Alpha-X Development
Developer/manufacturer of
electronics products & software
INFO: Product info including
electronics related software
EMAIL: info@alphax.com
HTTP://www.alphax.com/
RATING: (10)

Alpha Communications
Sound, intercoms, security & video
INFO: Electronic catalog, company
info
EMAIL: info@alpha-comm.com
HTTP://www.alpha-comm.com/
RATING: (8)

Alpha Metals
Manufacturer of solder materials
INFO: Product info, tech articles
EMAIL: n/a
HTTP://www.alphametals.com/
RATING: (8)

Alpha Technologies
Power supply manufacturer
INFO: Product info/specs/apps
EMAIL: alpha@alpha-usa.com
HTTP://www.alpha-us.com/
RATING: (8)

Alps Electric USA
Computer peripheral manufacturer
INFO: Tech support, contact,
drivers, FAQs
EMAIL: See site for current address
HTTP://www.alpsusa.com/
RATING: (9)

Alta Technology
Embedded computing
INFO: Product/software info/
support
EMAIL: sales,
support@altatech.com
HTTP://www.xmission.com/
~altatech/
RATING: (7)

Altec Lansing Corporation
Sound systems
INFO: Product info
EMAIL: markivaudiotech@qtm.net
HTTP://www.markivaudio.com/
altec/
RATING: (7)

Altera Corporation
Programmable logic manufacturer
INFO: Tech support, product
lookup
EMAIL: sos@altera.com (tech)
HTTP://www.altera.com/
FTP://ftp.altera.com/
RATING: (9)

Alternate Source Components Ltd.
Distributor or electronic component
parts
INFO: Company site
EMAIL: sales@alternatesrc.com
HTTP://www.alternatesrc.com/
RATING: (6)

Amateur Electronic Supply
Distributor of ham radio equipment
INFO: Locations, pricing, coupons, links
EMAIL: help@aesham.com
HTTP://www.aesham.com/
RATING: (9)

Amaze Electronics Corp.
Sells PC-based test equipment and PLCC prototype adapters
INFO: Product info
EMAIL: amaze@hooked.net
HTTP://www.hooked.net/users/amaze/
RATING: (7)

Amazon.com
Online bookstore
INFO: Select from 1 million book titles
EMAIL: See site for listing
HTTP://www.amazon.com/
RATING: (10)

AMCC (Applied Micro Circuits Corporation)
PCI and network interface ICs
INFO: Company site, tech support
EMAIL: pciinfo@amcc.com
HTTP://www.amcc.com/
RATING: (9)

America II Group
Includes America II Electronics, Inc.,
America II International,
America II Electronics,
A1 Teletronics, Inc.,
NICR, Inc.,
Mammoth Memory, Inc.,
America II Computer, Inc.,
Graveyard Electronics,
America II West, Inc.,
Q-1 Technologies, Inc.,

AMX PLC,
America II Direct
INFO: Links to their company sites
EMAIL: see individual sites for Emails
MAIN HTTP://www.america2.com/
RATING: (10)

American Arium
Manufactures emulators
INFO: Product info, support, links
EMAIL: info@arium.com
HTTP://www.arium.com/
FTP://ftp.arium.com/pub/
RATING: (8)

American Electronics Association Credit Union
Credit Union
INFO: Links, financial info, company info
EMAIL: info@aeacu.com
HTTP://www.aeacu.com/
RATING: (7)

American IC Exchange
Purchases and sells ICs
INFO: Tons of IC info including market pricing
EMAIL: webstaff@aice.com
HTTP://www.aice.com/
RATING: (9)

American Megatrends
BIOS, PC board manufacturer
INFO: Tech, FAQs, support, you name it!
EMAIL: See site for listing
HTTP://www.megatrends.com/
FTP://ftp.megatrends.com/ftp/
RATING: (9)

A

American Microsystems, Inc. (AMI)
Semiconductors
INFO: Products, apps, case studies, support
EMAIL: webmaster@poci.amis.com
HTTP://www.amis.com/
RATING: (8)

American Power Conversion Corp. (APC)
Power protection products manufacturer
INFO: Product info/support, help files, contests, you name it!
EMAIL: apcinfo@apcc.com
HTTP://www.apcc.com
RATING: (10)

American Precision Industries
Delevan/SMD divisions
Inductor solutions
INFO: Company site
EMAIL: apisales@delevan.com
HTTP://www.delevan.com/
RATING: (6)

American Zettler, Inc.
LCDs, LEDs, relays, etc., manufacturer
INFO: Products, sales
EMAIL: azettler@attmail.com
HTTP://eemonline.com/zettler/
RATING: (7)

Americomm
Sourcing supplier for electronics components
INFO: Company site
EMAIL: peter@ameri-comm.com
HTTP://www.ameri-comm.com/
RATING: (6)

Ameritech, Inc.
Telecommunication products/services
INFO: Products, news, links
EMAIL: webmaster@www.ameritech.com.
HTTP://www.ameritech.com/
RATING: (8)

Amitron
Thick-film tech manufacturer
INFO: Resistor info
EMAIL: kritson@amitron.com
HTTP://www.amitron.com/
RATING: (6)

AMP, Incorporated
Interconnection devices and systems manufacturer
INFO: Product and service, online catalog, support, company info, etc.
EMAIL: product.info@amp.com
HTTP://www.amp.com/
RATING: (10)

AMP on-line catalog
Online AMP catalog
INFO: Register to use 70,000 parts catalog
EMAIL: See site for listing
HTTP://connect.amp.com/
RATING: (9)

Amphenol Corporation
Interconnect product manufacturer
INFO: Product and company info
EMAIL: southard@amphenol.com
HTTP://www.amphenol.com
RATING: (7)

ANADIGICS, Inc.
Supplier of GaAs ICs
INFO: Company site
EMAIL: hr@anadigics.com
HTTP://www.anadigics.com/
RATING: (6)

Analog Devices, Inc.
IC manufacturer
INFO: Company site, tech support,
publications, software
EMAIL: See site for directory
HTTP://www.analog.com/
FTP://ftp.analog.com
RATING: (9)

Andraka Consulting Group
Digital hardware design
INFO: Papers, company info, links,
FPGA
EMAIL: randraka@ids.net
HTTP://www.ids.net/~randraka
RATING: (7)

Andrew Errington's PIC Project Page
Homepage
INFO: Projects for PICs, links, tons
of info
EMAIL:
a.errington@lancaster.ac.uk
HTTP://www.lancs.ac.uk/people/
cpaame/pic/pic.htm
RATING: (8)

Ansoft Corp.
Electromagnetic field simulation
software
INFO: Products, news, support,
requests
EMAIL: webmaster@ansoft.com
HTTP://www.ansoft.com/
RATING: (7)

Antec, Inc.
Computer parts supplier
INFO: Tech support, product info,
drivers, etc.
EMAIL: antec@antec-inc.com
HTTP://www.antec-inc.com/
FTP://ftp.antec-inc.com
RATING: (7)

APEM Components, Inc.
Switch components
INFO: Online quote, product info
EMAIL: info@apem.com
HTTP://www.apem.com/
RATING: (6)

Apple Computer, Inc.
Computer manufacturer
INFO: Product info/research/news/
development, you name it!
EMAIL: See site for listing
HTTP://www.apple.com/

Apple Computer Technical Library
HTTP://www.info.apple.com
RATING: (10)

Applied Laser Systems (ALS)
EMAIL: als@cdsnet.net

Applied Materials
Manufacturer of semiconductor
equipment
INFO: Products, technology, jobs
EMAIL: See site for forms
HTTP://www.appliedmaterials.com/
RATING: (9)

Applied Materials Etch Products
INFO: Products, news, overviews

A

EMAIL:
amat_prodinfo@amatetch.com
HTTP://www.amatetch.com/
RATING: (8)

Applied Microsystems Corporation
Embedded systems design, debug, test tools
INFO: Product info, resources, app notes, etc.
EMAIL: info@amc.com
HTTP://www.amc.com/
RATING: (9)

Applied Wave Research
Microwave, electromagnetic software/consulting
INFO: Company info/capabilities, free software
EMAIL: info@appwave.com
HTTP://www.appwave.com/
RATING: (7)

Apsuron Technologies, Inc.
Manufacturers' agent for OEMs
INFO: Connects to multiple company profiles
EMAIL: sales@apsuron.com
HTTP://www.apsuron.com/
RATING: (6)

Aptix Corp.
System emulation
INFO: News, company info, etc.
EMAIL: sales@aptix.com
HTTP://www.aptix.com/
RATING: (7)

ARC Electronics, Inc.
Industrial electronic parts, supplies, and engineering

INFO: Company site
EMAIL:
arcelect@uky.campus.mci.net
HTTP://www.arcelectronics.com/
RATING: (6)

Arcolectric Switches PLC
Switches/indicator lights manufacturer
INFO: Company info, contact
EMAIL: info@arcoswitch.co.uk
HTTP://www.arcoswitch.co.uk/
RATING: (4)

ARDUINI
Analog/power electronics/ consulting/products/ services
INFO: Tech papers, references, company site
EMAIL: doug@arduini.com
HTTP://www.arduini.com/
RATING: (9)

Argus Technologies DC Power Systems
Telecommunication power systems
INFO: Product info/specs, factory training, links
EMAIL:
alech@argus.iceonline.com
HTTP://www.argus.ca/argus/
RATING: (8)

Ariel Corporation
Expert in DSP, computer telephony, remote access servers, ADSL, and imaging
INFO: DSP hardware and software, high-density modems, ADSL technology, machine vision
EMAIL: ariel@ariel.com
HTTP://www.ariel.com/
RATING: (10)*****

Aries Electronics, Inc.
Electronic component manufacturer
INFO: Data sheets, help with products
EMAIL: info@arieselec.com
HTTP://www.arieselec.com/
RATING: (8)

Aromat Corporation
Manufactures NAiS Worldwide brand relays/switches/sensors/PLCs/lighting products/GPS/etc.
INFO: R&D, product info/sheets, engineer contact
EMAIL: See site for division contacts
HTTP://www.aromat.com/
RATING: (10)

AROW Components & Fasteners, Inc.
Electronics fastener distributor
INFO: Products, free spec handbook, a few links
EMAIL: arow@best.com
HTTP://www.arow.com/
HTTP://www.arowcom.com/
RATING: (7)

Arrick Robotics
PC-based motion control systems
INFO: Tons of PIC and robotic resources with forum for your projects. Well worth an hour of your surf time.
EMAIL: info@robotics.com
HTTP://www.robotics.com/
RATING: (10) *****

Arvid Technology Ltd.
Canadian motherboard & video products design/manufacturing
INFO: Info request, drivers, product info
EMAIL: info@arvida.ca
HTTP://www.arvida.ca/
RATING: (8)

AST Computer
Computer manufacturer
INFO: Products, service, support, news
EMAIL: webmaster@ast.com
HTTP://www.ast.com/
RATING: (10)

Astec America, Inc.
Power supply manufacturer
INFO: Company site, app notes
EMAIL: n/a
HTTP://www.astec.com/
RATING: (9)

ASUSTeK Computer, Inc.
Motherboard manufacturer
INFO: Full product info/support/news
EMAIL: webmaster@asustek.asus.com.tw
HTTP://www.asus.com.tw/
FTP://ftp.asus.com.tw/pub/
RATING: (10)

AT&T
Telecommunications company
INFO: You name it!
EMAIL: See site for directory
HTTP://www.att.com/
HTTP://www.research.att.com/
RATING: (10)

Atanas Parashkevov's Home Page
INFO: Links to real-time tech.
EMAIL: ata@cs.adelaide.edu.au

A

HTTP://www.cs.adelaide.edu.au/
users/ata/
RATING: (6)

Atec, Inc.
Flexible engineering/manufacturing
INFO: Outsourcing, newsletter, RF
link info
EMAIL: flex@atec.com
HTTP://www.sccsi.com/atec/
home.html
RATING: (9)

Atek Electronics, Inc.
Custom cable assemblies
INFO: Contact, company info
EMAIL: info-atek@atekelec.com
HTTP://www.atekelec.com/
RATING: (3)

ATI Cahn (Analytical Technology, Inc.)
Instrumentation manufacturer
INFO: Product info/specs, forum
EMAIL: See site for listing
HTTP://netopia.com:80/aticahn/
RATING: (8)

ATI Technologies, Inc.
Graphics/multimedia boards
INFO: Products, drivers,
developers, prices
EMAIL:
76004.3656@compuserve.com
HTTP://www.atitech.ca/
RATING: (9)

Atlanta Signal Processors, Inc. (ASPI)
DSP solutions for voice and audio
systems
INFO: Product info, software info

EMAIL: info@aspi.com
HTTP://www.aspi.com
BBS: 404-892-3200
RATING: (7)

Atlantic Components (Atcom)
Distributor of passives/electro-
mechanical
INFO: Linecard, tech data
EMAIL: See site for form
HTTP://atcomhq.com/
RATING: (7)

Atlas/Soundolier
Loudspeakers
INFO: Product info, specs
EMAIL: dane@sysint.com.
HTTP://www.sysint.com/systems/
atlas/
RATING: (7)

Atmel
NV-memory and logic circuits
manufacturer
INFO: Product info/support, news,
jobs
EMAIL: webleads@atmel.com
HTTP://www.atmel.com/
RATING: (9)

Austral Control Systems, Inc. (ACS)
Industrial/technical training
courses
INFO: Courses offered, schedules,
register
EMAIL: jtaylor@acsystems.com
HTTP://www.dworld.com/acs/
RATING: (7)

AutoCap Solutions, Inc.
Smart battery backup/management
systems
INFO: Product info/papers/specs &
other helpful info

EMAIL: See site for form
HTTP://www.autocap.com/
RATING: (7)

Autodesk, Inc. (AutoCAD)
Computer-aided drafting software
INFO: Product info, news,
education
EMAIL: webmaster@autodesk.com
HTTP://www.autodesk.com/
RATING: (10)

Automata, Inc.
Fabricates printed wire boards
INFO: Reviews, services, Q.A.
manual
EMAIL: webmaster@automata.com
HTTP://www.automata.com/
RATING: (8)

Automated Computer Technology, Inc. (ATC)
Independent in-circuit testing of
loaded PCBs and test engineering
services
INFO: Service info, customer list
EMAIL: webmaster@acthost.com
HTTP://www.acthost.com/
RATING: (6)

Automated Lifestyle
Home automation in TX
INFO: Home automation info
EMAIL: mavrick6@ix.netcom.com
HTTP://www.geocities.com/
Heartland/2841/
automatedlifestyle.html
RATING: (4)

Autotest Company
Testers for the power product
industry
INFO: Hardware, software info,
news
EMAIL: See site for forms & listing

HTTP://www.autotest.com/
RATING: (7)

Avance Logic, Inc.
Graphics and video semiconductors
INFO: Product, news, drivers,
EMAIL: sales@avance.com
HTTP://www.avance.com/
RATING: (8)

Avanti Electronics, Inc.
Electronics distributor
INFO: Linecard, contact
EMAIL: avantiel@cuy.net
HTTP://www.arsdata.com/avanti/
RATING: (5)

Avex Electronics, Inc.
Electronics manufacturing/
engineering
INFO: Technology, industry,
commerce sections. Great info on
electronics manufacturing
technology.
EMAIL: wwwavex@huber.com
HTTP://www.huber.com/Avex/
avex95/avexmain.htm
RATING: (9)

Avista Design Systems
Analog/RF/microwave design
software
INFO: News, notes, links, free
software
EMAIL: avista@avista.com
HTTP://www.avista.com/
RATING: (7)

AVM Technology, Inc.
PC audio products
INFO: Apex, Summit, Summit SST
MIDI/digital recording modules

A - B

EMAIL: sales@avmtechnology.com
HTTP://www.avmtechnology.com
RATING: (9)

Avnet (Corporate)
Components & products principally
for industrial customers
INFO: Their multiple companies'
links
EMAIL: See individual sites for
addresses
HTTP://www.avnet.com/
RATING: (7)

AVO International
Test and measuring instruments for
electrical power
INFO: Newsletter, tech notes, etc.
EMAIL: See site for form
HTTP://www.avointl.com/
RATING: (9)

Avtech Electrosystems Ltd.
Manufactures pulse/waveform/etc.,
generators and components
INFO: Product selection/
description/prices/apps
EMAIL: info@avtechpulse.com.
HTTP://www.avtechpulse.com/
RATING: (8)

AVX Corp.
Electronic components
INFO: Product info, tech support
software
EMAIL: avx@avxcorp.com
HTTP://www.avxcorp.com/
RATING: (8)

Award Software International, Inc.
BIOS, PC cards for several products

INFO: News, liturature, techsheets,
BIOS data
EMAIL: support@award.com
HTTP://www.award.com/
RATING: (7)

B.G. Micro, Inc.
Surplus ICs and electronics
INFO: Catalog, new products,
contact
EMAIL: bgmicro@ix.netcom.com
HTTP://www.bgmicro.com/
RATING: (8)

B. Richards & Associates
PCB design service
INFO: Company info, hot list of
links
EMAIL: brich@starcon.com
HTTP://www.starcon.com/bririch/
RATING: (7)

B&B Electronics
Serial communications tools
INFO: Data sheets, library, custom
design, links, software
EMAIL: bfranklin@bb-elec.com
HTTP://www.bb-elec.com/
FTP:// ftp://ftp.bb-elec.com/bb-elec/
RATING: (9)

Baradine Products Ltd.
Design/development of electronic
devices
INFO: Company site, a few links
and related data
EMAIL: baradine@stargate.ca
HTTP://www.baradine.com/
baradine/
RATING: (6)

Barnett Electronics, Inc.
Buys/sells used repeaters, etc.
INFO: Product available
EMAIL: johnb2@barnettelec.com

HTTP://www.barnettelec.com/
RATING: (3)

Basler Electric
Manufacturers of voltage
regulators, relays, etc.
INFO: Divisional product info/
apps/charts
EMAIL: webmaster@basler.com
HTTP://www.basler.com/
RATING: (7)

Batteries Plus
Misc. batteries for sale in TX
INFO: Company site
EMAIL: battery@onramp.net
HTTP://rampages.onramp.net/
~battery/
RATING: (5)

Battery Barn
Rechargeable batteries and
accessory sales
INFO: Product info
EMAIL: mpoint@batterybarn.com
HTTP://www.batterybarn.com/
RATING: (6)

BCD Electronics
Purchase and resale of surplus
electronics
INFO: Stock list, buy list, links
EMAIL: bcdelect@onramp.net
HTTP://www.bcdelectro.com/
RATING: (8)

B.E.A.M Robotics — Homepage
World of autonomous living
machines
INFO: Links, info about BEAM
robots, games, etc.
EMAIL: kolivas@lanl.gov
HTTP://sst.lanl.gov/robot/

HTTP://www.webconn.com/~mwd/
beam.html
RATING: (9)

Beau Tech
Electronics development tools
INFO: Catalog, distributor listing
with some links
EMAIL: info@beautech.com
HTTP://www.beautech.com/
RATING: (7)

Beckman Instruments, Inc.
Instrument manufacturer
INFO: Extensive papers on
specialized instruments
EMAIL: See site for form
HTTP://www.beckman.com/
RATING: (10)

Beginners' PIC Page
Hobbyist page for PIC apps
INFO: Tons of helpful info for PIC
projects. Also check out Matt's
homepage for addition links.
EMAIL: matt@atlas.kingston.ac.uk
HTTP://hobbes.king.ac.uk:80/matt/
pic/
RATINGS: (9)

Beige Bag Software
Circuit and design software
INFO: B2 Spice demos, links
EMAIL: info@beigebag.com
HTTP://www.beigebag.com/
FTP://beigebag.com/pub/vendor/
beigebag/
RATINGS: (8)

Belden Wire and Cable
HTTP://www.industry.net/
belden.wire

B

Bell Electronics Test & Measurement Equipment
Buys and sells used equipment
INFO: Company info, specials
EMAIL: sales@bellelect.com
HTTP://www.bellelect.com
RATING: (6)

Bell Industries
Manufactures electronics, computer graphics
INFO: Parts inquiry, tech support, links
EMAIL: info@bellind.com
HTTP://www.bellind.com/
RATING: (7)

The Bell Jar — Steve Hansen
Journal for vacuum technique
INFO: Links, indices to articles, you name it!
EMAIL: shansen@tiac.net
HTTP://www.tiac.net/users/shansen/belljar/
RATING: (9)

Bell Microproducts
Huge electronics distributor
INFO: You name it!
EMAIL: See site for directory
HTTP://www.bellmicro.com
RATING: (10)

Bellgraphics
Analog IC design
INFO: ASIC design service info
EMAIL: bell@bellgraphics.com
HTTP://www.bellgraphics.com/
RATING: (3)

BEMA
SEE: Electronics Guide

Benchmarq Microelectronics, Inc.
Manufactures ICs/modules for power management
INFO: Product info, data book, news
EMAIL: See site for map/listing
HTTP://www.benchmarq.com/
RATING: (9)

Berg Electronics, Inc.
Manufacturer of interconnectors
INFO: Product and company information
EMAIL: webmaster@bergelect.com
HTTP://www.bergelect.com/
RATING: (6)

Bertan High Voltage Corp.
High voltage power supplies
INFO: Product/company info/support/specs
EMAIL: info@bertan.com
HTTP://www.li.net/~bertan/
RATING: (7)

Bet Technology, Inc.
Developers and distributors of proprietary casino games
INFO: General information on games
EMAIL: bobk@betech.com
HTTP://www.betech.com/
RATING: (8)

Bill's Business and Industry Hotlist
Electronics links
EMAIL: wdbishop@dictator.uwaterloo.ca
HTTP://www.pads.uwaterloo.ca/~wdbishop/business.html
RATING: (9)

Bill's CB & 2-Way Radio Service
2-way, CB, marine, and GMRS radio sales & service
INFO: Online catalog
EMAIL: info@bills2way.com
HTTP://www.bills2way.com
RATING: (8)

BittWare Research Systems
DSP hardware
INFO: Product info/datasheets, specs
EMAIL: info@bittware.com
HTTP://www.bittware.com
FTP://ftp.bittware.com
RATING: (9)

Bivar
Electronic hardware/optoelectronic assemblies
INFO: Product info, design own LED assembly, catalog
EMAIL: bivar@interserv.com
HTTP://www.bivar.com/
RATING: (9)

BizWeb
Web directory including electronics
INFO: Links to tons of electronics sites
EMAIL: bob@bizweb.com
HTTP://www.bizweb.com/keylists/electronics.html
RATING: (9)

BJM Electronics, Ltd.
Network/cabling/audio/video products provider
INFO: Linecard
EMAIL: bjm@dorsai.org
HTTP://www.bjm.com/
RATING: (3)

Bliley Electric Company
Crystals and oscillators
INFO: Product info, search, help
EMAIL: info@bliley.com
HTTP:// www.bliley.com/
RATING: (8)

Blue Fin Technologies
OEM sales and purchasing
INFO: Jobs, Blue Fin group links
EMAIL: icsales@bluefin.com
HTTP://www.bluefin.com/
RATING: (8)

Blue Sky Research
TeX software
INFO: Product info/sales/support, TeX stuff
EMAIL: bluesky@ix.netcom.com
HTTP://www.bluesky.com/
RATING: (7)

Bob Colyard's Shortwave & DXing Page
Shortwave radio hobby information page
INFO: List of frequencies, tips, news, links
EMAIL: slapshot@cybercomm.net
HTTP://www.cybercomm.net/~slapshot/speedx.html
RATING: (9)

Bodine Electric Company
Manufacturer of motion control and motors
INFO: Company products, literature
EMAIL: bodine19@starnetinc.com
HTTP://www.bodinemtr.com/
RATING: (7)

B

Borland International, Inc.
C, C++, Pascal, etc., development software
INFO: Programs, seminars, news, demos, forums
EMAIL: See site for form
HTTP://www.borland.com/
FTP://see www for access
RATING: (10)

Bourns
Pressure transducers products
HTTP://www.industry.net/bourns

Brady
HTTP://www.industry.net/bradyipd

Brandenburg
High-voltage power supply vendor
INFO: Catalog, new products, support
EMAIL: brandenburg@astec.co.uk
HTTP://www.brandenburg.co.uk/
RATING: (8)

Brett Bymaster's Circuit and Software
INFO: Great hobbyist site with tons of projects!
EMAIL: n/a
HTTP://expert.cc.purdue.edu/~bymas/
RATING: (10)

Brigar Electronics
Sells component parts, power supplies, etc.
INFO: Specials, prices
EMAIL: brigar@binghamton.com
HTTP://www.binghamton.com/brigar/
RATING: (6)

Broadband Communication Products, Inc.
Manufacturers of fiberoptic communication/test equipment
INFO: Product news/specs/info, links
EMAIL: fiberlink@bcpinc.com
HTTP://www.bcpinc.com
RATING: (7)

Brockton Electronics Ltd.
Manufacturer — Cable and wire harnesses
INFO: Company info and products
EMAIL: info@brockton-electronics.com
HTTP://www.brockton-electronics.com/
RATING: (7)

Brookdale Electronics, Inc.
Crystal/oscillator distributor
INFO: Linecard, catalog, crystal search
EMAIL: info@brookdale.com
HTTP://www.brookdale.com
RATING: (7)

Brooktree Corp.
Digital and mixed-signal solutions
INFO: Tech data, schematics, software, company info
EMAIL: Form on site or webmaster@brooktree.com
HTTP://www.brooktree.com/
FTP://ftp.brooktree.com/pub/apps/
RATING: (9)

Brother Industries, Ltd.
Consumer electronics, printers, etc.
INFO: Product info/support, jobs
EMAIL: webmaster@brother.co.jp
HTTP://www.brother.com/
RATING: (9)

Bryston Ltd.
Manufactures specialty audio
electronics
INFO: Products, warranty, reviews
EMAIL: bwrussell@bryston.ca
HTTP://www.bryston.ca/
RATING: (8)

Bud Industries, Inc.
Enclosures/cabinets manufacturer
INFO: Custom info, service, reps
EMAIL: bud@starlink.com
HTTP://www.budind.com/
RATING: (7)

Bumper Industries
High fidelity loudspeakers
INFO: Simple company page
EMAIL: bumper@ipof.fla.net
HTTP://www.fla.net/bumper/
home.html
RATING: (6)

Burisch Elektronik Bauteile
Electronic components distributor,
in Austria
INFO: Product info, links
EMAIL: info@beb.usa.com
HTTP://ourworld.compuserve.com/
homepages/beb_com/
RATING: (8)

Burr-Brown Corp.
Signal-processing ICs
INFO: Articles, tech info, company
info
EMAIL: See site for directory
HTTP://www.burr-brown.com/
RATING: (8)

Bussmann Division — Cooper Industries
Fuses for electronics and power
apps

INFO: Company & product info,
catalog.
EMAIL: fusebox@bussmann.com
HTTP://www.bussmann.com/
RATING: (10)

Butterfly DSP, Inc.
High-end DSP chips/modules
INFO: DSP apps/FAQs/design/
studies
EMAIL: mflemin@pacifier.com
(VP of Engineering)
HTTP://www.butterflydsp.com/
RATING: (8)

Butterworth-Heinemann
Publishes books, including
electronics
INFO: Brief on all books they
publish
EMAIL: info@repp.com
HTTP://www.bh.com/
RATING: (8)

Byte Craft Limited
Software tools for embedded
systems
INFO: Product spec, tips, new,
demos
EMAIL: info@bytecraft.com
HTTP://www.bytecraft.com
RATING: (8)

C & K Components, Inc.
Miniature switches, trimmers, etc.
INFO: Company/product info,
catalog ordering
EMAIL: info@ckcorp.com
HTTP://www.ckcorp.com/
RATING: (8)

C

C-Cube Microsystems
ICs for digital video
INFO: Product/company/tech info
EMAIL: webmaster@c-cube.com
HTTP://www.c-cube.com/
RATING: (9)

C Z Labs
Computer/telephone/cable
connectors
INFO: Online catalog, tech info,
etc.
EMAIL: czlabs@rcknet.com
HTTP://www.czlabs.com/
RATING: (7)

C-MAC Industries, Inc.
Microelectronic modules, crystals,
etc.
INFO: Company info, news and
contacts
EMAIL: See site for directory
HTTP://www.cmac.ca/
RATING: (7)

C.P. Clare Corp.
Semiconductors, magnetic switchs/
relays
INFO: Company info, apps, news
EMAIL: suggestions@cpclare.com
HTTP://www.cpclare.com/
RATING: (8)

Cable Systems Intl.
Communications products
INFO: Company info/contacts, tech,
huge apps.
EMAIL: See site for directory
HTTP://www.csi-cables.com/
RATING: (9)

CAD Related Internet Sites
By Information Management
Technologies
INFO: Links to every CAD resource
on the NET
EMAIL: See IMTs page
HTTP://www.webcom.com/~imt/
wwwcad.html
RATING: (10)

Cadence Design Systems, Inc.
Automated design and CAD
software company
INFO: 'Plugged-In Magazine'
includes features, interviews, etc.
EMAIL: See site for form
HTTP://www.cadence.com/
RATING: (10)

Caig Laboratories, Inc.
Enviro-friendly solder supplies
INFO: FAQs, links, data sheets
EMAIL: caig123@aol.com
HTTP://www.caig.com/
RATING: (8)

California Eastern Labs
Agent for NEC Semiconductors,
etc.
INFO: Locations, jobs, contacts,
tech data
EMAIL: cwilhoit@cel.com
HTTP://www.cel.com/
RATING: (7)

California Switch & Signal, Inc.
Electromechanical components
distributor
INFO: Links, company info/contact
EMAIL:
mark@csmain.calswitch.com
HTTP://www.calswitch.com/
RATING: (10)

CAM RPC Electronics
Electronic component distributor
INFO: Linecard, locations
EMAIL: n/a
HTTP://www.camrpc.com/
electronics/
RATING: (5)

Canadian Circuits, Inc.
PCB development
INFO: Company site
EMAIL: cancirc-info@starcon.com
HTTP://www.starcon.com/cancirc/
RATING: (6)

Canare Cable
Manufactures cable/connectors/
patchbays
INFO: Online library/catalog,
specials
EMAIL: canare@canare.com
HTTP://www.canare.com/
RATING: (8)

CanaREP, Inc.
Electronics company representative
in CA
INFO: Company info, linecard
EMAIL: See site for listing
HTTP://www.canarep.com/
RATING: (6)

Capilano Computing Systems, Ltd.
Electronics design tools software
INFO: Demos, links, support, you
name it!
EMAIL: info@capilano.com
HTTP://www.capilano.com/
FTP://ftp.wimsey.com/pub/
capilano/
RATING: (9)

Capital Electronics
Design/manufacturing of PCB
INFO: Shareware PCM design
programs
EMAIL: sales@capital-elec.com
info@capital-elec.com for
electronic brochure
HTTP://www.capital-elec.com/
RATING: (9)

Capital Electronics, Inc.
Distributor of electronic
components
INFO: Stock, company info,
linecard
EMAIL: ceisales@capelec.com
HTTP://www.capelec.com/
RATING: (9)

Cardinal Components, Inc.
Xtals and oscillators
INFO: Product data and news
EMAIL:
cardinal@cardinalxtal.com
HTTP://www.cnct.com/home/
cardinal/
RATING: (8)

Carlton-Bates
Industrial electronic distributor
INFO: Online inventory, company
data
EMAIL: See site for area listings
HTTP://www.carlton-bates.com/
RATING: (10)

Carnegie Mellon University Robotics Club
INFO: Links, projects, club info,
etc.
EMAIL: rc99@andrew.cmu.edu

C

HTTP://
www.contrib.andrew.cmu.edu/usr/
rc99/rc.html
RATING: (8)

Carroll Touch, Inc.
BBS: 512-388-5668

Cascade Design Automation Corp.
Automation software for IC design
INFO: News, product info/support
EMAIL: info@cdac.com
HTTP://www.cdac.com/
RATING: (8)

Case Amateur Radio Club
INFO: Projects, links, etc.
EMAIL: sct@po.cwru.edu
HTTP://cnswww.cns.cwru.edu/
misc/w8edu/
RATING: (7)

CAST — Computer Aided Software Technologies, Inc.
VHDL design software/training
INFO: Models, datasheets, training info
EMAIL: info@cast-inc.com
HTTP://www.cast-inc.com/
RATING: (9)

CeeJay
Distributor of reed relays, optos, etc.
INFO: Line info/specs/schematics
EMAIL: sales@ceejay.com
HTTP://www.ceejay.com/
RATING: (7)

Central Resources, Inc.
Database components, assemblies, products. Mfgr/Distr. homepages with individual databases.
EMAIL:
webmaster@centralres.com
HTTP://centralres.com/
FTP:///ftp.centralres.com/
RATING: (7)

CFX Group, Inc.
Distributor of inter-connects
INFO: Company site/info, pricing
EMAIL: sales@cfxgroup.com
HTTP://www.cfxgroup.com/
RATING: (7)

Channel System Components
Distributes electronic components
INFO: Company info, links
EMAIL: info@channelsys.com.sg
HTTP://www.channelsys.com.sg/
RATING: (8)

Cherry Electric
Switches, sensors, displays
HTTP://www.industry.net/
cherry.electrical

Cherry Semiconductor Corp.
EMAIL: info@cherry-semi.com
HTTP://www.cherry-semi.com/
Business card

Chesapeake Musical Equipment
Tube amp repair
EMAIL: loizcren@erols.com

Chip Directory
Jaap van Ganswijk's page
INFO: A listing of almost every chip imaginable, links to related sites and manufacturers and FAQ

library. Extremely valuable site for everyone in electronics.
EMAIL: ganswijk@xs4all.nl
HTTP://www.hitex.com/chipdir
HTTP://www.xs4all.nl/~ganswijk/chipdir/
Mailing List:
majorodomo@amc.com, body: subscribe chipdir-1
RATING: (10) *****

Chip Express
ASICs
INFO: Articles, news, software, data sheets.
EMAIL: moreinfo@chipx.com
HTTP://www.chipexpress.com/
RATING: (10)

Chip Supply
Semiconductor die distributor
INFO: Linecard, newsletter, jobs
EMAIL: rmarshall@chipsupply.com
HTTP://www.chipsupply.com/
RATING: (7)

Chips & Technologies, Inc.
Semiconductor and software — ASICs
INFO: Data sheets, corp. info and drivers
EMAIL: See site for departments
HTTP://www.chips.com/
RATING: (7)

Chromatic Research
MPACT media processors
Mediaware/Multimedia processors
INFO: Company/product info, news, jobs, media processor background info
EMAIL: info@chromatic.com

HTTP://www.chromatic.com/
HTTP://www.mpact.com/
RATING: (8)

Chrysalis Symbolic Design, Inc.
EDA software
INFO: Product info, news, seminar info
EMAIL: www@chrysalis.com
HTTP://www.chrysalis.com/
RATING: (7)

Chuck Schwark's
Antique radio resource page
INFO: Resources/links for restorers/collectors, Philco info
EMAIL: caschwark@aol.com
HTTP://members.aol.com/caschwark/
RATING: (9)

Circuit Assembly Corp.
Connectors
INFO: Product data, company info
EMAIL: See site for form
HTTP://www.ca-online.com/
RATING: (7)

Circuit Cellar Ink & Circuit Cellar, Inc.
Magazine publisher & hardware manufacturer
INFO: Links, magazine info, home control system details
EMAIL: info@circellar.com
HTTP://www.circellar.com/
FTP://ftp.circellar.com/pub/circellar/
RATING: (8)

C

Circuit Cookbook WWW Page
EE student homepage
INFO: A myriad of circuits and
software to download
EMAIL: charro@ee.ualberta.ca
HTTP://www.ee.ualberta.ca/
~charro/cookbook/
RATING: (10)

Circuit Repair Corp.
Board repair and rework
INFO: Free offers, great handbook
on repairing PCBs, and tons of
other repair notes
EMAIL: info@circuitnet.com
HTTP://www.circuitnet.com/
RATING: (10)*****

Circuit Specialists, Inc.
Electronics parts catalog
INFO: Online catalog
EMAIL: See site for form
HTTP://www.cir.com/
RATING: (10)

Circuit-Test Electronics Ltd.
Electronics equipment/product
manufacturer & distributor
INFO: Product line with pics
EMAIL: info@cirtest.com
HTTP://www.cirtest.com/
RATING: (8)

Cirrus Logic — Crystal Semiconductor — PCSI
Multimedia, communications, data
storage
INFO: Links, company info,
support
EMAIL: See site for listings
HTTP://www.cirrus.com/
RATING: (10)

Clairtronic
Mains, transformers, AC/DC
adaptors
INFO: UK order form
EMAIL: sales@clairtronic.co.uk
HTTP://www.clairtronic.co.uk/
RATING: (6)

CMP Enclosures
Electronic enclosures & accessories
INFO: Complete photos and specs
EMAIL: See site for form
HTTP://www.enclosures.com/
RATING: (8)

CMP's TechWeb
Technology super site with online
magazines
Also see individual sites:
— Computer Reseller News
— Electronic Buyer's News
— Electronic Engineering Times
— OEM Magazine
— VAR Business
INFO: Online magazines
EMAIL: See site for form
HTTP://www.techweb.com/
RATING: (10)*****

Coast Air, Inc.
Electronic components & hardware
distributor
EMAIL: coastair@earthlink.net for
more info
HTTP://home.earthlink.net/
~coastair/
RATING: (3)

Coastal Concepts
Vacuum tubes and electronic
components
INFO: Weekly updated listing of
thousands of vacuum tubes for
misc. equipment

EMAIL: tekman@execpc.com
HTTP://www.execpc.com/~tekman/
audio.html
HTTP://www.execpc.com/~tekman/
parts.html
RATING: (7)

Coilcraft
Inductors, transformers, filters
INFO: Great product info
EMAIL: info@coilcraft.com
HTTP://www.coilcraft.com/
RATING: (8)

Collins Printed Circuits
Rockwell — PCBs for avionics
INFO: Links, facilities, products,
services
EMAIL: See site for departments
HTTP://www.rockwell.com/
rockwell/bus_units/cca/cpc/
RATING: (10)

COMDEX Online
Computer/electronics convention
INFO: Show info, questions,
register, etc.
EMAIL: staff@comdex.com
HTTP://www.comdex.com:8008/
RATING: (10)

Commodity Components Intl.
(CCI)
Distributor, brokering, surplus
INFO: Catalog, company info
EMAIL: cci@cci-inc.com
HTTP://www.cci-inc.com/
RATING: (7)

COMMtronics Engineering
Telecommunications engineering,
etc.
INFO: Control scanners from a
computer! Links, info, etc.
EMAIL: ccheek, bcheek@cts.com

C

HTTP://ourworld.compuserve.com/
homepages/bcheek/
FTP://ftp.cts.com/pub/bcheek
RATING: (9)

Comp.Robotics.Research
Newsgroup
Academic, government & industry
reseach in robotics
INFO: Archive of the newsgroup
since its beginning in May 1995
EMAIL: crr@robot.ireq.ca
(submissions)
crr-request@robot.ireq.ca
(information
HTTP://www.robot.ireq.ca/CRR/
RATING: (n/a)

Compaq Computer Corp.
Computer manufacturer
INFO: News, products, service/
support
EMAIL: support@compaq.com
HTTP://www.compaq.com/
FTP://ftp.compaq.com/
RATING: (10)

Component Connections
Kitting, hard-to-find electronic
components
INFO: Search engine, contact

EMAIL: See site for contacts
HTTP://
www.componentconnection.com/
RATING: (6)

Component Distributors, Inc.
Distributor
INFO: Products, links, design
support, shows

C

EMAIL: webmaster@compdist.com
HTTP://www.compdist.com/
RATING: (8)

Component Resources
Distributor of components, fasteners
INFO: Linecard, buying tips
EMAIL: christyc@compres.com
HTTP://www.compres.com/
RATING: (7)

Component Sales, Inc.
Distributor of electronic components
INFO: Linecard, order, check order
EMAIL: tarling@sonic.net
HTTP://www.azoff.com/
RATING: (7)

Compudigital Industries
Buys/Sells equipment, ICs, etc.
INFO: Stock, links
EMAIL: jeff_hilliard@compudigital.com
HTTP://www.compudigital.com/
RATING: (7)

Computer City
SEE: Tandy Corp.

Computer Reseller News
Newspaper for value-added reselling
INFO: Articles, news, stock quotes, etc.
EMAIL: See site for form
HTTP://techweb.cmn.com/crn/current/
RATING: (9)

Computing & Electrical Engineering
Heriot-Watt University Library
INFO: Plenty of links to keep you going for hours
EMAIL: libhelp@hw.ac.uk
HTTP://www.hw.ac.uk/libWWW/faculty/faceng/comele.html
RATING: (9)

Condel Technology
IC test sockets
INFO: Notes, product pics & specs
EMAIL: See site for listing
HTTP://www.condel.com/
RATING: (7)

Condor DC Power Supplies, Inc.
INFO: Product info, order catalog
EMAIL: condordc@condorpower.com
HTTP://www.fishnet.net/~condordc/
RATING: (8)

Consulting Specifying Engineer Online (CS-E)
Integrated engineers home on the web
INFO: Articles, issues, links, etc.
EMAIL: See site for editor directory
HTTP://www.csemag.com/
RATING: (8)

Consumer Electronics Manufacturers Association (CEMA)(EIA)
SEE ALSO: *Electronic Industries Association*
INFO: Consumer electronics news and information; info on Consumer Electronics Show (CES)
EMAIL: edwardk@eia.org
HTTP://www.eia.org/cema

HTTP://www.cemacity.org/ (by Jan. 1, 1997)
RATING: (9)

Control Design Supply, Inc.
Components distributor
INFO: Ordering info, product descriptions
EMAIL: sales@controldesignsupply.com
HTTP://www.controldesign.com/
RATING: (7)

Corcom, Inc.
Filters manufacturer
INFO: Catalogs, charts, FAQs, tips
EMAIL: info@cor.com
HTTP://www.cor.com/
RATING: (9)

Core Concepts
Publishes "PC Buyer's Handbook"
INFO: Book info, regular computer help articles
EMAIL: gforeman@poboxes.com
HTTP://www.netcom.com/ ~gforeman/small.html
RATING: (7)

Corning Electronics, Inc.
Distributor of electronics/computer products
INFO: Linecard, links
EMAIL: cornelec@vivanet.com
HTTP://www.vivanet.com/ ~cornelec/
RATING: (7)

Creative Labs, Inc.
Sound cards and other peripherals
INFO: Awesome multimedia company site!
EMAIL: See site for forms
HTTP://www.creaf.com/
RATING: (10)

C

Cree Research, Inc. & Real Color Displays (RCD)
SiC semiconductors
INFO: Silicon carbide info & company site
EMAIL: See site for directory
HTTP://www.cree.com/
RATING: (10)

Crosspoint Solutions, Inc.
EMAIL: info@xpoint.com

Crystaloid
EMAIL: crystaloid@aol.com

CTS Corp.
Variety of components for OEMs
INFO: News, products, data sheets
EMAIL: See site for directory
HTTP://www.ctscorp.com/reps.html
RATING: (8)

Custom Security Designs
Electronic kits, security devices, telephone devices
INFO: Plans for kits
EMAIL: csd@mindspring.com
HTTP://www.mingspring.com/ ~jtaylor
RATING: (9)

Cypress Electronics
Distributor of cooling fans, switches
INFO: Product info, contact
EMAIL: info@cypelex.com
HTTP://www.cypelex.com/
RATING: (7)

Cypress Online (Quarterly Newsletter)
EMAIL: sze@cypress.com

C - D

Cypress Semiconductor Corp.
Manufacturer of a multitude of semiconductors
INFO: Company site with everything you need to know about Cypress products. Includes apps, data sheets, software.
EMAIL: See site for directory
BBS: 408-943-2954
HTTP://www.cypress.com/
RATING: (9)

Cyrix Corp.
Designs/manufactures CPUs
INFO: FAQs, resellers directory, new product info
EMAIL: tech_support@cyrix.com
BBS: 214-994-8610
HTTP://www.cyrix.com/
FTP://www.cyrix.com/
RATING: (10)

D.Huras Software
Home automation software
INFO: X-10 CP 290 shareware, links
EMAIL: dhuras@inforamp.net
HTTP://ourworld.compuserve.com/ homepages/davidhuras
FTP://members.aol.com/ davidhuras/sharware
RATING: (9)

D.R.Components, Inc.
Components for Industry
INFO: Linecard, stock database
EMAIL: sales@drcomponents.com
HTTP://www.drcomponents.com/
RATING: (7)

D&D Battery, Inc.
Sells batteries for consumer electronics
INFO: Catalog with prices, company site
EMAIL: battery@4batteries.com
HTTP://www.4batteries.com/
RATING: (6)

D&M Electronics
Distributor of electronics/misc. products
INFO: Product info, order form, links
EMAIL: webguy@d-m.com
HTTP://www.d-m.com/
RATING: (8)

Daburn Electronics & Cable Corp.
INFO: Product info, helpful hints using wire, cable and shrink tubing
EMAIL: daburn@daburn.com
HTTP://www.daburn.com/~daburn/
RATING: (7)

Daitron, Inc.
Engineering & electronic systems
INFO: Company site
EMAIL: info@daitron.com
HTTP://www.daitron.com/
RATING: (5)

Dale Electronics, Inc.
Xtals, displays, resistors, caps, etc.
INFO: Product info/specs
EMAIL: n/a
HTTP://vishay.com/vishay/dale
RATING: (7)

Dalis Electronics, Inc.
Component distributor
INFO: Links, product line
EMAIL: dalis1@aol.com

HTTP://www.farrsite.com/dalis/
RATING: (7)

Dallas Semiconductor Corp.
IC manufacturer
INFO: Extensive product info/data/apps
EMAIL: See site for directory
HTTP://www.dalsemi.com/
FTP://ftp.dalsemi.com/pub/
RATING: (9)

Damac Products, Inc.
Enclosures, racks, etc.
INFO: Product info, helpful files
EMAIL: meberhart@damac.com
HTTP://www.damac.com/
RATING: (9)

Danari International
Buys/sells old products
INFO: Company linecard
EMAIL: d_nachman@prodigy.com
HTTP://www.themarketplace.com/wanttobuy/Danari.html
RATING: (5)

Daniel Woodhead Company
Manufactures electronics products
INFO: Product info with pics
EMAIL: dwinfo@interaccess.com
HTTP://www.danielwoodhead.com/
RATING: (7)

Dannell Electronics, Ltd.
UK distributor
INFO: Company info and catalog req.
EMAIL: sales@dannell.co.uk
HTTP://clever.net/qms/dannell.htm
RATING: (6)

Data I/O Corp.
EDA software/hardware
INFO: sales, products, tech info

D

EMAIL: webmaster@data-io.com
HTTP://www.data-io.com/
RATING: (9)

Data Institute Business School
Technical training school
INFO: Contains tutorials on many electronics/computer subjects
EMAIL: fgolden@nai.net
HTTP://datainstitute.com/
RATING: (10)

Data Instruments, Inc.
HTTP://www.industry.net/data.instruments

Dataforth Corp.
Industrial signal conditioning and data communications
INFO: Product info/news/support, links
EMAIL: techinfo@dataforth.com
HTTP://www.dataforth.com/
RATING: (9)

Datalogic, Inc.
Manufactures automatic ID products
INFO: News, product info
EMAIL: info@datalogic.com
HTTP://www.datalogic.com/
RATING: (7)

Dataman Programmers, Inc. (UK, USA, CA)
Device programmers
INFO: Product info, software, links. Well done site!
EMAIL: info@dataman.com
HTTP://www.dataman.com/
RATING: (10) ***

D

Davenham Satellite Systems
Supply of satellite TV spares & repairs
INFO: FAQs spares, books, accessories. UK satelitte treasures
EMAIL: davsat@netcentral.co.uk
HTTP://www.netcentral.co.uk/~davsat/
RATING: (8)

David B. Thomas Pic Programs
INFO: DTMF, MIDI sender, other projects
EMAIL: dthomas@rt66.com
HTTP://www.rt66.com/dthomas/pic/pic.html
RATING: (7)

David Tait's PIC Links and PIC Archive
INFO: PIC info, links, programmers, you name it!
EMAIL: david.tait@man.ac.uk
HTTP://www.man.ac.uk/~mbhstdj/piclinks.html
HTTP://www.man.ac.uk/~mbhstdj/files/
RATING: (9)

Davis Instruments
Test instruments
HTTP://www.industry.net/davis.instruments

DB ELECTRONICS
INFO: Help pages, contact
EMAIL: 73373.1051@compuserve.com
HTTP://ourworld.compuserve.com/homepages/db_electronics/
RATING: (8)

D.B. Roberts Company
Distributes electronic fasteners
INFO: Products, locations, links
EMAIL: sales@dbroberts.com
HTTP://www.dbroberts.com/
RATING: (8)

DC Battery Specialists
Wholesale distributor of batteries
INFO: Battery FAQ, product info
EMAIL: sales@dcbattery.com
HTTP://www.dcbattery.com/
RATING: (9)

DC-Tech
Reconditioned test and measurement
INFO: Catalog, Internet services
EMAIL: dctech@california.com
HTTP://www.dctech.com/
RATING: (7)

DCI (Digital Communications, Inc.)
Intermods
INFO: Product and helpful files with reviews
EMAIL: dctech@california.com
HTTP://www.dci.ca/
RATING: (9)

Debco Electronics, Inc.
Electronics parts house/custom computer builder
INFO: Online catalog, bargains, etc.
EMAIL: debc@debco.com
HTTP://www.debco.com/
RATING: (9)

Dedicated Devices
System level automation modules
INFO: SLAM info, products/apps/demo/lit.

EMAIL: jongabay@li.net
HTTP://www.li.net/~jongabay/
slam.html
RATING: (9)

Defiant Eurosat
Satellite and other resources
INFO: Satellite TV hacks, FAQs,
links. Worth a long look!
EMAIL: defiant@eurosat.com
HTTP://www.eurosat.com
RATING: (10) *****

Delco Electronics Corp.
Division of Hughes Electric
Automotive electronics
INFO: Tech papers, apps, trivia,
links
EMAIL: See site for listing
HTTP://www.delco.com/
RATING: (10)

Delevan
See: *American Precision Industries*

Dell Computer Corp.
Computer manufacturer
INFO: Product info/support,
purchase
EMAIL: See site for form
HTTP://www.dell.com/
RATING: (10)

Designer's Den
PCB designers page by George
Patrick
INFO: PCB design resources and
info
EMAIL: gpatrick@aracnet.com
HTTP://www.aracnet.com/
~gpatrick/
RATING: (9)

Dialogic Corp.
Telephony hardware/software
INFO: Tips, products, locations,
support
EMAIL: sales@dialogic.com
HTTP://www.dialogic.com/
RATING: (10)

Digi-Key Corp.
Catalog parts house
INFO: Catalog, contact, many links
EMAIL: webmaster@digikey.com
HTTP://www.digikey.com
RATING: (10)

Digital Equipment Corp. (DEC)
Semiconductor division
Digital equipment manufacturer
INFO: Huge company site with
library of products, support files,
etc.
EMAIL: See site for listing
HTTP://www.digital.com/info/
semiconductor
FTP://ftp.digital.com//pub/digital/
info/semiconductor/
RATING: (10)

Digital Equipment Corp. (DEC)
Semiconductor operations
HTTP://www.dec.com/
HTTP://www.digital.com/
HTTP://www.digital.com/info/
semiconductor
EMAIL:
semiconductor@digital.com
EMAIL: moreinfo@digital.com

Digital Market, Inc.
Open sourcing system
INFO: Sourcing site info, links

D

EMAIL:
support@digitalmarket.com,
HTTP://www.digitalmarket.com/
RATING: (7)

Digital Semiconductor
See Digital Equipment Corp.
HTTP://www.digital.com/info/
semiconductor

Digital Storage Scope.FAQ
By: John D. Seney
EMAIL: john@wd1v.mv.com
HTTP://www.mv.com/ipusers/
wd1v/dsofaq.html
RATING: (8)

Digital Vision, Inc.
Home of ComputerEyes frame
grabbers and TelevEyes scan
converters
INFO: Product information, FAQs,
sample images, notable quotables,
etc.
EMAIL: webmaster@digvis.com
HTTP://www.digvis.com/
RATING: (10)

The Dilbert Zone
Computer humor! ONLINE
COMIC STRIP
INFO: Recent strip, Scott Adams
info, etc.
EMAIL: scottadams@aol.com,
webmaster@unitedmedia.com
HTTP://www.unitedmedia.com/
comics/dilbert/
RATING: (10)

Diodes Incorporated
Mfg. of discrete semiconductors

INFO: Product summaries with data
sheets
EMAIL: diodes-info@diodes.com
HTTP://www.diodes.com/
RATING: (7)

Disel USA, Inc.
Component distributor
INFO: Linecard
EMAIL: disel@xroads.com
HTTP://www.xroads.com/~disel/
RATING: (4)

DiskDude's Virtual Page
Homepage
INFO: Game console, PC info/
hacks. Great site!
EMAIL: diskdude@poboxes.com
HTTP://internex.net.au/~papa/
RATING: (10)

Distinct Micro, Inc. (DMI)
Sourcing company
INFO: Inventory, articles, links
EMAIL: bruce@dmi.qc.ca.
HTTP://www.dmi.qc.ca/
RATING: (8)

Diverse Electronics
Components distributor
INFO: Quotes, line info/pics
EMAIL: diverse@odyssee.net
HTTP://www.odyssee.net/~diverse/
RATING: (7)

DIY Electronics
Electronic Kits
INFO: Links, helpful files, product
info
EMAIL: diykit@hk.super.net
HTTP://www.hk.super.net/~diykit/
RATING: (9)

Don Lancaster
SEE: *The Guru's Lair*

Don's Corner
Satellite, TV, uC programming
INFO: Home page with links
EMAIL: n/a
HTTP://helix.net/~lekei/
RATING: (8)

Don's Workshop - DonTronics Home Page
Low-cost DIY PCB kits
INFO: A treasure chest of uC info, links
EMAIL: dontronics@labyrinth.net.au
HTTP://www.labyrinth.net.au/~donmck/
RATING: (10)

Donald L. Klipstein, Jr.
Hobbyist page
INFO: Led/laser/high voltage/etc. A great site with links and useful data.
EMAIL: don@misty.com
HTTP://www.misty.com/~don/
RATING: (9)

Douglas Electronics
EDA software for MacOS, stock and custom PCB manufacturing
INFO: Product/service info
EMAIL: info@douglas.com
HTTP://www.douglas.com/
RATING: (7)

Douglas W. Jones — U of Iowa
Control of stepping motors tutorial
INFO: Tutorial with schematics, links
EMAIL: jones@cs.uiowa.edu
HTTP://www.cs.uiowa.edu/~jones/step/
RATING: (9)

D

Dr. Dobb's Website
Magazine publisher
INFO: Issues, source code, links, tons of info!
EMAIL: editors@ddj.com
HTTP://www.ddj.com/
RATING: (9)

DSP Development Corporation
Markets DADiSP, software for engineers
INFO: Demo sotware, info, free student edition
EMAIL: info@dadisp.com
HTTP://www.dadisp.com/
RATING: (8)

DSP Group, Inc.
DSP applications developer
INFO: Telephony equipment products, news, etc.
EMAIL: webmaster@dspg.com
HTTP://www.dspg.com/
RATING: (8)

DSP Tools, Inc.
DSP development boards, software
INFO: Hardware/software, services
EMAIL: jervisd@dsptools.com
HTTP://www.dsptools.com/
RATING: (8)

dTb Software
EDA and CAE software for engineers
INFO: Demos, product info, examples
EMAIL: info@dtbsware.com
HTTP://www.dtbsware.com/
RATING: (8)

D - E

Duracell USA
Battery manufacturer
INFO: New products, tips, etc.
EMAIL: See site for form
HTTP://www.duracell.com/
RATING: (10)

DY 4 Systems, Ltd.
VME solutions manufacturer
INFO: Product/company info,
glossary
EMAIL: support@dy4.com
HTTP://www.dy4.com/
RATING: (10)

Dycor Industrial Research, Inc.
Data acquisition system
development, etc.
INFO: Data aquisition product info.
EMAIL: sales@dycor.com
HTTP://www.dycor.com/
RATING: (8)

Dyna Art Designs
Instant PCBs using specialty paper
and a laser printer
INFO: Product info, help files,
links, refs
EMAIL: dynaart@netport.com
HTTP://www.dynaart.com/
RATING: (9)

Dynamic Hybrids, Inc.
Engineering of MCMs, etc.
INFO: Product info
EMAIL: dyhybrid@localnet.com
HTTP://www.ieec.binghamton.edu/
ieec/ieecpdyn.html
RATING: (3)

E-LAB Digital Engineering, Inc.
Embedded single-board computers
and semi-custom integrated circuits
INFO: Product info, application
notes, comprehensive electronics
resource. Directory with links.
EMAIL: elab@netins.net
HTTP://www.netins.net/showcase/
elab/
RATING: (10)

E-Mark, Inc.
Interconnects importer/distributor
INFO: Products, reps, catalog
request
EMAIL: bwest@e-markinc.com
HTTP://www.e-markinc.com/
RATING: (7)

E-Switch
INFO: Online catalog, company
info
EMAIL: info@e-switch.com
HTTP://www.e-switch.com/
RATING: (8)

E-T-A Circuit Breakers
Circuit breakers for equipment
INFO: Product info/data
EMAIL: See site for country list
HTTP://www.etacbe.com/
RATING: (7)

E2W3 — by Full Circle Connections, Inc.
Electrical engineering on the World
Wide Web
INFO: Every possible EE resource
at a mouse click!
EMAIL: See site for form
HTTP://www.e2w3.com/
RATING: (10) **********

EAC Electronics Company
Reed relays
INFO: Company/product info
EMAIL: reedman@gnn.com
HTTP://www.reedrelay.com/
RATING: (7)

Eagle Design Automation, Inc.
Embedded systems hardware/
software
INFO: Product info, news
EMAIL: sales@eagledes.com
HTTP://www.eagledes.com/
RATING: (7)

EAO Switch Corp.
Switch supplier
INFO: Company info, product info
EMAIL: info@eaoswitch.com
HTTP://www.eaoswitch.com/
RATING: (7)

Earthmind, Inc.
Electronics, software, mechanical
devices
INFO: Product info, pricing
EMAIL: earthmind@worldlink.ca
HTTP://www.worldlink.ca/~earth
RATING: (6)

Eastern Acoustic Works
Stage and movie sound equipment
INFO: Idea exchange, news,
products, plenty of visual treats.
EMAIL: info@eaw.com
HTTP://www.eaw.com/
RATING: (10)

Eaton Corp.
Supplier of a multitude of products
INFO: List of subsidiaries and
product names
EMAIL: corpcomm@eaton.com
HTTP://www.eaton.com/
RATING: (8)

ebm Papst
Fans, blowers for electronics
INFO: Company site
EMAIL: sales@ebm.com
HTTP://www.ebm.com/
RATING: (4)

EBN Online (Electronic Buyers' News)
Online electronics magazine
INFO: Archives, calendar, you
name it!
EMAIL: rclapso@cmp.com
HTTP://techweb.cmp.com/ebn/
current/default.html
RATING: (10)

Echelon Corp.
Control networks
INFO: Backgrounds, data books,
tech info
EMAIL: See site for listing
HTTP://
www.lonworks.echelon.com/
RATING: (9)

Ecliptek Corp.
Crystals, oscillators, and inductors
INFO: Product/company info, price
request
EMAIL: ecsales@ ecliptek.com
HTTP://www.ecliptek.com/
RATING: (7)

Economy 2-Way Distributors, Inc.
Wholesale distributor
INFO: Linecard
EMAIL: econ2way@aol.com
HTTP://members.aol.com/
econ2way/
RATING: (3)

E

ECSC EIO WWW Server
Computers and electronics surplus
INFO: Forums, catalog, specials, links. A true information site!
EMAIL: ecsc@caprica.com, ecsc@eio.com
HTTP://www.eio.com/
RATING: (10)

EDAC, Inc.
Connectors
INFO: Product/company info
EMAIL: edac@edac.net
HTTP://www.edac.net/
RATING: (6)

EDCO, Inc., of Florida
UL listed surge suppressors
INFO: Product info, bulletins
EMAIL: edco@mercury.net
HTTP://www.edosurge.com/
RATING: (8)

Eddy Electronics
Printed circuit fabrication
INFO: Enviro-friendly PCB info
EMAIL: mwo@freenet.edmonton.ab.ca
HTTP://www.freenet.edmonton.ab.ca/eddymfg
RATING: (7)

Edinburgh Engineering Virtual Library (EEVL)
THE gateway to engineering information in the UK
INFO: Searchable database containing descriptions of many electronics sites
EMAIL: webmaster@eevl.icbl.hw.ac.uk
HTTP://www.eevl.ac.uk/
RATING: (9)

EDN Access — Magazine
The design information source of the electronics industry
INFO: Cover story, subscribe, software
EMAIL: See site for directory
HTTP://www.ednmag.com/
RATING: (9)

EE/CS Mother Site
Maintained by Nabeel Ibrahim
INFO: Over 800 links to companies, organizations, universities, etc.
EMAIL: ibrahim@leland.stanford.edu
HTTP://www-ee.stanford.edu/soe/ieee/eesites.html
RATING: (10)

EE Design Magazine
SEE: *Integrated Systems Design*
HTTP://www.eedesign.com/

EE Times on the Web
Online electrical engineering magazine
INFO: Archives, news, links, FAQs, forums, etc.
EMAIL: tmoran@cmp.com
HTTP://www.eet.com/
RATING: (10)

EEM Online
SEE: *Electronic Engineers Master*

EENet, Inc.
The electronics industry inforum & interactive workplace
INFO: Just about everything to do with electronics
EMAIL: campbell@eenet.com

E

Efficient Networks, Inc.
ATM adapters/software
INFO: Product/company info. Great help page for ATM info!
EMAIL: webmaster@efficient.com
HTTP://www.efficient.com/
RATING: (9)

EG3 Communications
Electronic engineering Net resources
INFO: Electronic Engineers' Toolbox, Virtual Conference, Electronics Search FAQ. Top-level engineering site!
EMAIL: inquiry@eg3.com
HTTP://www.cera2.com
RATING: (10)**********

Eimac Div — CPI
Radio/power grid tubes
INFO: Products, purchase, radio club
EMAIL: See site for form
HTTP://www.eimac.com/
RATING: (8)

EJE Research
Electronic contract manufacturing
INFO: Outsourcing info, manufacturing services
EMAIL: ejemail@eje.com
HTTP://www.eje.com/
RATING: (9)

Elanix, Inc.
Electronics CAD/simulator software
INFO: Product eval, newletter, apps, FAQs
EMAIL: elanix@elanix.com

HTTP://www.elanix.com/
RATING: (9)

Elantec Semiconductor
Analog ICs
INFO: Product info, links, datasheets
EMAIL: webmaster@elantec.com
HTTP://www.elantec.com/
RATING: (8)

Electric Switches
HTTP://www.industry.net/
electric.switches

Electrical and Electronic Engineering —EEE-Alert
Signal processing, image/speech comm.
INFO: Online magazines
EMAIL: m.eligh@elsevier.nl
HTTP://www.elsevier.nl:80/eee/
RATING: (8)

Electrical Engineering Circuits Archive
Jerry Russell's site
INFO: Circuits, datasheets, models, uPs, text files, software,
links, Q-n-A
EMAIL: pfloyd@u.washington.edu
HTTP://www.ee.washington.edu/
eeca/
RATING: (10)*****

ElectriTek & Subsidiaries
Distributes batteries, passive devices, displays
INFO: Company/product info
EMAIL: jgordow@intalt.com

E

HTTP://www2.csn.net/~jgordow/
eltkhp.html
RATING: (7)

Electro-Mech
Manufactures switches/indicator
lights
INFO: Product info, sales
EMAIL:
75013.1645@compuserve.com
HTTP://ourworld.compuserve.com/
homepages/electro_Mech/
RATING: (5)

Electro-Technology Corp.
Spare parts/service to magnetic
recording industry
INFO: Parts, services, sales
EMAIL: info@electro-
technology.com
HTTP://www.electro-
technology.com/
RATING: (7)

Electrochem Lithium Batteries
A product of Wilson Greatbatch,
Inc.
INFO: Product info, sales contact
EMAIL: tklem@greatbatch.com
HTTP://www.greatbatch.com
RATING: (7)

Electroglas, Inc.
Supplier of automated wafer-
probing products
INFO: Products, stock quote,
locations
EMAIL:
marketing@electroglas.com
HTTP://www.electroglas.com/
RATING: (9)

Electron Tubes, Inc.
EMAIL: phototubes@aol.com

Electronic Article Surveillance
Homepage by Ted Park
INFO: Explanation of electronic
security tags
EMAIL: tpark@beer.org
HTTP://www.beer.org:80/~tpark/
knogomag.html
RATING: (8)

Electronic Bulk Buys —
Wirz Electronics Sponsored
Site
Custom electronics design
INFO: FAQs, Catalog, specials, you
name it!
EMAIL: blw2@cec.wustl.edu
HTTP://cec.wustl.edu/~blw2/
index.html
RATING: (10) *****

Electronic Components &
Equipment, Ltd.
Australian large-quantity
distributor
INFO: Company info, parts search
EMAIL: info@ece.com.au
HTTP://centralres.com/cgi-win/
eceltd.exe
RATING: (5)

Electronic Countermeasures,
Inc.
Cellular, Fax & paging monitoring
systems
INFO: Specs, demo, owners manual
EMAIL: bill.fischer@t8000.com
HTTP://www.t8000.com/eci/eci.htm
RATING: (7)

Electronic Distributors
Amateur radio, wideband receivers

INFO: Product info/specs, dealers, suggest pricing
EMAIL: edco@vni.net
HTTP://www.elecdist.com
RATING: (8)

Electronic Energy Control, Inc.
Relay interface for connection to RS232
INFO: Catalog with pics, sounds, links
EMAIL: eeci1@ibm.net
HTTP://www.eeci.com/
RATING: (10)

Electronic Engineers Master — EEM Online
Company listings site
INFO: EEM online, EEM local resources, *Electronic Products* magazine, etc.
EMAIL: See site for form
HTTP://www.eemonline.com/
RATING: (10)

Electronic Engineering Times
See: *EE Times*

Electronic Engineers' Toolbox
See: *EG3 Communications*

Electronic Equipment Company
Parts distributor
INFO: Products, news, locations
EMAIL: See site for listing
HTTP://www.eeco.com/
RATING: (7)

Electronic Industries Association (EIA)
SEE ALSO: Consumer Electronics Manufacturers Assoc.
INFO: Association links/info/etc.

EMAIL: publicaffairs@eia.org.
HTTP://www.eia.org/
RATING: (Over-all: 10)

Electronic Information
Mega-Links page by Tor Kristjan Berge
INFO: Links to FTP, FAQs,WWW, Misc.
EMAIL: tkb@fagmed.uit.no
HTTP://ev-www.ser.fm.uit.no/ electronics/index.html
RATING: (10)

Electronic Innovations
Custom circuit design
INFO: Services offered, links
EMAIL: jkaeh@aol.com
HTTP://www.uso.com/eic/eic.html
RATING: (4)

Electronic Maintenance Supply Company (EMSCO)
Distributor of various components, equipment
INFO: Specials, order info, product news
EMAIL: See site for form
HTTP://www.hammondelec.com/ emspage.html
RATING: (6)

Electronic Measurements, Inc.
Power supplies manufacturer
INFO: Product info
EMAIL: See site for form
HTTP://www.emipower.com/
RATING: (7)

Electronic News On_Line
Online electronics-related newpaper
INFO: Current, back issues

E

EMAIL: enews@sumnet.com
HTTP://www.sumnet.com/enews/
RATING: (10)

Electronic Parts Supply
Computer, industial electronics
distributor
INFO: Photo catalog, pricing
EMAIL: wondarl@hooked.net
HTTP://www.entrepreneurs.net/
wondarl/eps.htm
RATING: (8)

Electronic Products
SEE ALSO: *EEM ONLINE*
The engineer's magazine of product
technology
INFO: Online mag with plenty of
info
EMAIL: See EEM site for form
HTTP://
www.electronicproducts.com/
RATING: (10)

Electronic Products &
Technology (EP&T)
Canadian electronics industry
magazine
INFO: EP&T online, subscribe to
mag.
EMAIL: info@ept.ca
HTTP://www.ept.ca/
RATING: (9)

Electronic Professional
Services, Inc.
Electronic engineering firm
INFO: Products, services, links,
news
EMAIL: mrf@eps1.com
HTTP://www.apk.net/eps
RATING: (10)

Electronic Systems, Inc.
Supplier to OEMs only
INFO: Industry, company info,
contact
EMAIL: sales@iw.net
HTTP://www.iw.net/electsys
RATING: (5)

Electronic Technology Corp.
(ETC)
Mixed-signal ICs manufacturer
INFO: ASICs, standard product
info/specs, etc.
EMAIL:
webmaster@etechcorp.com
HTTP://www.etechcorp.com/
RATING: (7)

Electronic Tool Company, Inc.
(ETCO)
Precision electronics tools
distributor
INFO: Links, contact
EMAIL: ecto@computer.net
HTTP://www.circuitworld.com/
etco/homepa~1.htm
RATING: (7)

Electronics & Electrical
Engineering Laboratory
National Institute of Standards and
Technology
INFO: Projects, publications,
software, reference, you name it!
EMAIL: eeel@nist.gov
HTTP://www.eeel.nist.gov/
RATING: (10) *****

Electronics Component News
Online
Magazine publishers site —
Chilton Company
INFO: Industry phonebook,
subscriptions, etc.

EMAIL: See site for forms
HTTP://www.ecnmag.com/
RATING: (9)

The Electronics Guide — BEMA sponsored
FTP and World Wide Web sites for microcontrollers
INFO: Huge links to semiconductor, uC, etc. sites
EMAIL: bema@worldaccess. nl
HTTP://www.worldaccess.nl/~bema/
RATING: (10) *****

The Electronics HomeWorld
Homepage of Vincent Himpe
INFO: Just about everything related to electronics is on this site!
EMAIL: Vincent.Himpe@ping.be
HTTP://www.ping.be/~ping0751/
RATING: (10)

Electronics Information Files
Diana Todd homepage
INFO: Schematics, hobby help files, great for electronics hobbyists
EMAIL: diana@texas.net
HTTP://lonestar.texas.net/~diana/electro.htm
FTP://members.aol.com/schematix/
RATING: (9)

Electronics Information Online (EIO)
See: *ESCS*

Electronics Links
CyberLepak Corner
INFO: Basic Stamp and other electronics links
EMAIL: Site has form
HTTP://www.geocities.com/SiliconValley/3319/electronics.html
RATING: (8)

Electronics Manufacturers on the NET
Directory of electronics hardware products on the Net
INFO: Manufacturers' links, add a link. Search can be product or manufacturer
EMAIL: elx@webscope.com
HTTP://www.webscope.com/elx/homepage.html
RATING: (10)

The Electronics Manufacturing Guide
Guide to all facets of electronics manufacturing
INFO: Links to manufacturers, suppliers, online mag, etc.
EMAIL: cirworld@norwich.net
HTTP://www.circuitworld.com/
RATING: (10)

Electronics on the Web
Bimonthly online magazine
INFO: Past and present issues. Great articles for hobbyists!
EMAIL: nickh@emags.u-net.com
HTTP://www.emags.com/electron.htm
RATING: (9)

Electronics Search FAQ
SEE: *EG3 Communications*

Electronics Source
Links provided by Miller Enterprises
INFO: Massive OEM, distributor and news links
EMAIL: sponsor@electsource.com
HTTP://www.electsource.com/
RATING: (10) ***

E

Electronics Workbench
Interactive Image Technology, Ltd.
Electronics simulation software
INFO: Demo, news, links
EMAIL: ewb@interactiv.com
HTTP://www.interactiv.com/
FTP://ftp.interactiv.com
RATING: (9)

Electronix Corp.
Instructional video, service
software, & electronic parts
INFO: A repair technician's
goldmine of links, tips, downloads,
part search, you name it!
EMAIL: sales@electronix.com
HTTP://www.electronix.com/
RATING: (10) ***********

Electronix Express
Mail order electronics supplies for
schools, industry
INFO: Equipment, components,
kits, specials, tech tips
EMAIL: electron@elexp.com
HTTP://www.elexp.com/
RATING: (9)

Electroswitch Electronic Products
Rotary switches and
electromechanical design solutions
INFO: Company, design service
info
EMAIL: sales@electro-nc.com
HTTP://www.electro-nc.com/
RATING: (6)

Electrotek Concepts, Inc.
Engineering consulting
INFO: Huge information site for
power quality issues

EMAIL: See site for directory
HTTP://www.electrotek.com/
FTP://ftp.electrotek.com/pub/
RATING: (10)

Electrotex
Broad-line electronic parts
distributor
INFO: News, specials, www
services
EMAIL: See site for form
HTTP://www.electrotex.com/
electrotex.html
RATING: (9)

Elephant Memory Computer Memory Reseller
INFO: Links, prices, RAM
explanations
EMAIL: bryon@master.net
HTTP://www.master.net/elephant
RATING: (7)

Elma Electronics, Inc.
Enclosures, components
INFO: Tech briefs, product info,
reps
EMAIL: sales@elma.com
HTTP://www.elma.com/
RATING: (9)

Elmec Technology of America
MCI Mail: 539-2234

Elo Touchsystems, Inc.
Kiosks, touch browsers and other
Touchscreen products
INFO: Touchscreen monitors,
software, tech, libraries, etc,
EMAIL: eloinfo@elotouch.com
HTTP://www.elotouch.com/
RATING: (10) ***

Elsag Bailey Process Automation
Automation solutions
INFO: Locations and links to Divisions
EMAIL: webmaster@bailey.com
HTTP://www.bailey.com/
RATING: (7)

Embedded Systems Internet Resources
Homepage of Charlie Daly
INFO: Tons of uC data and links
EMAIL:cdaly@compapp.dcu.ie
HTTP://www.compapp.dcu.ie/
~cdaly/embed/
RATING: (9)

Embedded Systems Programming
Embedded.com and MWMedia-merged megasite
INFO: Magazine, code, resources, jobs, vendors, links, you name it!
EMAIL: See site for listing
HTTP://www.embedded.com/
RATING: (10) *****

EMCO High Voltage Company
EMAIL: emco@ix.netcom.com

EME Fan & Motor, Inc. /Sunon
AC/DC fans manufacturer
INFO: Product/company info/specs
EMAIL:
104065.3051@compuServe.com
HTTP://www.sunon-eme.com/
RATING: (7)

Emergency Help Line
Components distributor
INFO: Office locations, hot list
EMAIL: sales@ehl.com
HTTP://www.ehl.com/
RATING: (6)

Emulation Technology
Interconnectors/adaptors/emulators
INFO: Product info with pics
EMAIL:
webmaster@pmail.emulation.com
HTTP://www.emulation.com/
RATING: (7)

Endicott Research Group, Inc. (ERG)
Power supplies for information displays
INFO: Apps, datasheets, parts catalog
EMAIL: johnpete@worldnet.att.net
HTTP://www.ergpower.com/
RATING: (8)

Engineering Information, Inc.
Engineering Information Village
INFO: 30 day free trial, demo database
EMAIL: See site for listing
HTTP://www.ei.org/
RATING: (9)

Enterprise Electronics Corp.
Adapters, transformers, misc. parts
INFO: Product info, links, tech papers
EMAIL: sales@entcorp.com
HTTP://www.entcorp.com/
RATING: (9)

Entest
Test equipment distributor/rep
INFO: Linecard, contact
EMAIL:
73504.2127@compuserve.com
HTTP://www.themetro.com/01/
entest/
RATING: (5)

E

EPI Motion Systems
Servo Motion Control Products
INFO: PC/104 servo motor
controller, Turn-Key motion control
applications.
EMAIL: motion@epimbe.com
HTTP://www.epimbe.com/motion
RATING: (9)

Epson America, Inc.
Consumer electronics
INFO: Online tech support, links,
drivers, you name it!
EMAIL: See site for listing
HTTP://www.epson.com/
RATING: (9)

EPROM Types List
FTP://oak.oakland.edu/pub/misc/
eprom/eprom-types.list

Eric Electronics
Passives and misc. parts distributor
INFO: Product links, linecard,
contact
EMAIL: ee@ericnet.com,
irvine@ericnet.com
HTTP://www.ericnet.com/
RATING: (7)

Eric Smith's PIC Projects
Homepage
INFO: PIC projects/info galore!
EMAIL:
eric@goonsquad.spies.com
HTTP://www.brouhaha.com/
%7Eeric/pic/
RATING: (10)

Ericsson
Telecommunications, components
manufacturer
INFO: Libraries, product info/news/
apps/data/support
EMAIL: See site for local listings
HTTP://www.ericsson.se/
RATING: (9)

Esco-Insulectro
Distributes insulation, passives, etc.
INFO: Online ordering, links, etc.
EMAIL: See site for listing
HTTP://www.insulectro.com/
eihome.htm
RATING: (9)

Essex Group, Inc.
Wire and cable manufacturer
INFO: Products, technology, links
EMAIL:
essex_homepage@essexgroup.com
HTTP://www.essexgroup.com/
RATING: (7)

ETEQ Components Pte Ltd.
Distributes components, Singapore
INFO: Product listing, contact
EMAIL: eteqcomp@singnet.com.sg
HTTP://www.singnet.com.sg/
~eteqcomp/
RATING: (3)

Etronix Manufacturing Pte Ltd.
Contract manufacturing services to
electronics industry
INFO: Services offered, contact
EMAIL: etronix@singnet.com.sg
HTTP://www.singnet.com.sg/
~etronix
RATING: (7)

EURAS USA, Inc.
Information and support to repair technicians
INFO: Explanation of EURAS System
EMAIL: See site for form
HTTP://www.euras.com
RATING: (7)

The European Satellite Hack Page
SEE: *Defiant Eurosat*

Eurotherm and Divisions
Design/manufacture/sales of electronics equipment
INFO: Controls, drives, gauging systems, etc.
EMAIL: See site for form
HTTP://www.eurotherm.com/
RATING : (9)

Evacuated Envelopes
Online valve amp magazine
INFO: Tons of info on valve amps
EMAIL: tubes@hillier.demon.co.uk
HTTP://members.aol.com/valveaudio/index.html
RATING: (10)

Eveready
Battery manufacturer
INFO: Battery bunny bonanza!
EMAIL: See site for form
HTTP://www.eveready.com/
RATING: (8)

Everett Charles Technologies
Test fixtures/equipment, kits
INFO: Online catalog, news, apps, links
EMAIL: stongeg@ectinfo.com
HTTP://www.ectinfo.com/
RATING: (8)

Exar Corp.
Mixed-signal IC manufacturer
INFO: Articles, news, product info/apps
EMAIL: sales@exar.com
HTTP://www.exar.com/
RATING: (9)

Excel Electronics Supply
Electronic component distributor
INFO: Overview
EMAIL: excelint@ix.netcom.com
HTTP://www.silcom.com/excel/
RATING: (3)

Exetech Software Innovations (ESI)
Molecular Beam Epitaxy software
INFO: Software explanation, contact
EMAIL: exetech@prairienet.org
HTTP://www.prairienet.org/exetech/
RATING: (4)

Exide Battery Corporation
Battery manufacturer
INFO: Product guide, breakthroughs
EMAIL: See site for form
HTTP://www.exideworld.com/power/
RATING: (10)

EXPEER International, Inc.
Website servicing EE and product development industry
INFO: Forums, ref libraries, links, you name it!
EMAIL: See site for form
HTTP://www.expeer.com/
RATING: (9)

E - F

Exponential Technology, Inc.
Designs/builds microprocessors
INFO: Jobs, company info, news
EMAIL: info@exp.com
HTTP://www.exp.com/
RATING: (7)

Extech Instruments Corp.
Test equipment, data loggers, etc.
INFO: Product info, news, links,
software
EMAIL: extech@extech.com
HTTP://www.extech.com/
RATING: (8)

Extron Electronics
Computer-video interfacing
products mfgr.
INFO: Catalog, software, links
EMAIL: extrons3@attmail.com
HTTP://www.extron.com/
RATING: (10)

FAQ Library — Electronics
Library of known FAQs
INFO: Frequently asked questions
for a range of electronics subjects.
Should be one of your first
stopovers on the Web.
EMAIL: wwwadmin@faqlib.com
HTTP://www.faqlib.com/
electro.htm
RATING: (10) **********

Farnell
Catalog distributor of electronic
components
INFO: Semiconductor datasheets,
order info, links, you name it!
EMAIL: See site for info
HTTP://www.farnell.co.uk/
RATING: (9)

Fast Forward Engineering
Electronics design house
INFO: Q&A for uCs, links, product
info
EMAIL: fastfwd@ix.netcom.com
HTTP://www.geocities.com/
SiliconValley/2499/
RATING: (10)***

Federated Purchaser, Inc.
Electronics component and supplies
distributor
INFO: Linecard, products, links
EMAIL: fpi@worldnet.att.net
HTTP://
www.federatedpurchaser.com/
RATING: (7)

Ferrotronic Components, Inc.
Canadian distributor
INFO: Products carried, links
EMAIL: See site for listing
HTTP://www.accent.net/ferro/
RATING: (7)

Filip Gieszczykiewicz's Sites
TLJ consulting
Electronics, repair, auto, etc.
INFO: Extremely useful page to
technicians and electronics
hobbyists/professionals. Includeds
links, FAQs
EMAIL: filipg@paranoia.com,
kevintx@paranoia.com
HTTP://www.paranoia.com/~filipg/
RATING: (10) **********

Filip Gieszczykiewicz's Sites
Includes Sci.electronics FAQ:
magazines,
Sci.electronics FAQ: electronics-
oriented WWW links,
Sci.electronics.repair FAQ
SEE SITE FOR ADDITIONS

FineWare
Software geared to the shortwave hobbyist
INFO: Product info/support/ shareware, links
EMAIL: mfine@crosslink.net
HTTP://www.crosslink.net/~mfine/
RATING: (9)

First Magnitude Corp.
Manufactures low-cost slow-scan CCD cameras
INFO: Demo disk, CCD shots
EMAIL: firstmag@delphi.com
HTTP://www.wyoming.com/ ~johnson/fmc.htm
RATING: (6)

Fisher Audio/Video
Audio/video equipment manufacturer
INFO: News, contest, and of course products
EMAIL: See site for contact
HTTP://www.audvidfisher.com/
RATING: (10)

Flashpoint Technologies
Components distributor in UK
INFO: Pricing, offers, tech center
EMAIL: flashpoint@macro.co.uk
HTTP://www.inext.co.uk/ flashpoint/
RATING: (7)

Fluke Corporation
Electronics instruments
INFO: Products, apps, site has been under construction for some time
EMAIL: n/a
HTTP://www.fluke.com/
RATING: (6)

Fontronic Corp.
INFO: New-used test equipment
EMAIL: afoti@msn.com

Fordahl
Manufacturer of crystal oscillators (TCXOs, VC-TCXOs, VCXOs, OCXOs and PXOs), MCF and quartz crystals
INFO: Product/ company info
EMAIL: mnewcome@fordahl.com
HTTP://www.fordahl.com/
RATING: (7)

ForeSight Electronics, Inc.
Power products distributor
INFO: Product line, suppliers info
EMAIL: sales@fse-power.com
HTTP://www.fse-power.com
RATING: (8)

Fouraker Electronics
MRO electronics parts distributor
INFO: Location info
EMAIL: if3@nettally.com
HTTP://www.nettally.com/fouraker/
RATING: (3)

Forest Electronic Developments Home Page
PIC programmers/kits/software
INFO: Links, product info, news
EMAIL: gmwarner@lakewood.win-uk.net
HTTP://www.ibmpcug.co.uk/ ~gmwarner/fed.htm
RATING: (8)

Fox Electronics
Frequency control products manufacturer

F - G

INFO: Product data sheets/great site for crystals, oscillators, and more
EMAIL: sales@foxonline.com
HTTP://foxonline.com/
HTTP://fastfox.com
RATING: (9)

Fox Electronics

Oscillator products manufacturer
INFO: Products info/tech terms. Great info site for oscillators.
EMAIL: sales@foxonline.com
HTTP://foxonline.com/
RATING: (9)

Free Online Dictionary of Computing

Denis Howe's page
INFO: Search, links
EMAIL: dbh@doc.ic.ac.uk
HTTP://wombat.doc.ic.ac.uk/
RATING: (9)

FSI/FORK Standards, Inc.

Supplier of optical shaft encoders, photoelectric sensors
INFO: Under construction
EMAIL: info@fsinet.com
HTTP://www.xnet.com/~fsi
RATING: (3)

FTP for MCU/MPU Resources

8051, 68HC11, Z80, PIC, etc.
INFO: Compilers, schematics, you name it!
FTP://ftp.funet.fi/pub/microprocs/
RATING: (10)

Fujitsu America, Inc.

See site for affiliates

Manufacturer of various electronics products/components
INFO: Products, jobs, affiliates. Huge multiple sites with tons of info/datasheets
EMAIL: webmaster@fujitsu.com
HTTP://www.fujitsu.com/
RATING: (10)***

Fullman — Mechanical Contractors

Process piping for the semiconductor industry
INFO: Semiconductor production info, company info, links. Great info site!
EMAIL: sales@fullman.com
HTTP://www.fullman.com
RATING: (10)

Future Electronics

Component distributor in CA
INFO: Product links, company info, dealer listing
EMAIL: webmaster@future.ca
HTTP://www.future.ca/
RATING: (8)

Fuzzy Systems Engineering

Fuzzy logic tools & applications
INFO: Product, book, demos, papers, etc.
EMAIL: fuzzysys@electriciti.com
HTTP://www.fuzzysys.com/fuzzysys
RATING: (7)

G. Campbell and Associates (GCA)

Professional/technical services. Expertise in satellite communications, TT&C Earth Stations and automated test systems

INFO: List of company
capabilities, past contracts, contact
EMAIL:
100120.3702@compuserv.com
HTTP://www.wp.com/
mcintosh_page_o_stuff/gca.htm
RATING: (6)

Gateway Electronics
Full-line distibutor
INFO: Links, company info,
specials
EMAIL: gateway@mo.net
HTTP://www.gatewayelex.com/
RATING: (8)

GEC Electronics
Australian VAR distributor
INFO: Links, specials, etc.
EMAIL: davidm@gec.com.au/
HTTP://www.gec.com.au/
RATING: (10)

GEC Plessy Semiconductors
EMAIL: topasic@io.com

Gehrke Electronics
All types of electronic repair
INFO: Under construction
EMAIL:
electron@chatt.mindspring.com
HTTP://chatt.mindspring.com/
~electron
RATING: (n/a)

Genashor Corp.
EDA software and consulting
INFO: Products, services, tech, free
versions via ftp
EMAIL: genashor@pluto.njcc.com
HTTP://pluto.njcc.com/~genashor
FTP://pluto.njcc.com/pub/genashor
RATING: (8)

G

General Device Instuments (GDI)
IC device programmers
INFO: Products, software, support
— Great source for software
for programmers they stock.
EMAIL: icdevice@best.com
HTTP://www.generaldevice.com
RATING: (9) ***

General Electric Company (GE)
Consumer/military/business
products manufacturer/services
INFO: News, products, services,
Huge information site!
EMAIL: See site for forms
HTTP://www.ge.com/
RATING: (10)

General Instrument Corp.
Video, voice, data solutions
INFO: Product/company info, links
EMAIL: See site form
HTTP://www.gi.com/
RATING: (8)

Genesis Microchip
Video and imaging chips
INFO: Company/ product info,
datasheets
EMAIL: See site forms
HTTP://www.genesisus.com/
RATING: (8)

GENIE Interactive
Online service
Software, chat, video, forums, etc.
EMAIL: webmaster@gomax.com
HTTP://www.genie.com/
RATING: (10)

G

Gentron Corporation
Solid-state relays and power modules
INFO: Catalog ordering, new products, datasheets
EMAIL: request@gentron.com
HTTP://www.gentron.com/
RATING: (8)

Gerber Electronics
Distributor franchise
INFO: Specials, great online catalog with price, links
EMAIL: postmaster@gerberelec.com
HTTP://www.gerberelec.com/
RATING: (9)

Gernsback Publications, Inc.
Electronics Now, *Popular Electronics*
INFO: Content of recent 2 issues, discussion groups, index to articles, downloads, magazine services. Very hobbyist oriented!
EMAIL: See site for full listing
HTTP://www.gernsback.com/
FTP://ftp.gernsback.com/pub/
RATING: (9)

GHZ Technology
Supplier of silicon RF/microwave power transistors
INFO: Product info/sheets, reps
EMAIL: ghztch@alink.net
HTTP://www.ghz.com/
RATING: (8)

The Giant Internet IC Masturbator: Search Form
SEE ALSO: *Filip Gieszczykiewicz's Sites*

IC FAQ with pinouts, logic info, etc.
INFO: Richard Steven Walz's FAQ with a search form to search for pinouts and info on ICs
EMAIL: rstevew@armory.com, filipg@paranoia.com
HTTP://www.paranoia.com:80/~filipg/HTML/cgi-bin/giicm_form.html
RATING: (10)

Gisle's World
Homepage
INFO: PIC info, VF, satellite, TV
EMAIL: sting@mix.hsv.no
HTTP://mix.hsv.no/~sting/
RATING: (9)

GlobeTech International
Used automatic test equipment supplier
INFO: Tons of ATE info, links
EMAIL: info@gtinet.com
HTTP:www.gtinet.com/
RATING: (10)

Globtek, Inc.
Manufactures power supplies, cable, etc.
INFO: Products, catalog, company info
EMAIL: globtek1@chelsea.ios.com
HTTP://gramercy.ios.com/~globtek
RATING: (7)

Goldstar — LG Electronics, Inc.
Consumer electronics manufacturer
INFO: Products, support, jobs
EMAIL: See site for form
HTTP://www.goldstar.co.kr/
RATING: (7)

Golledge Electronics
Specialist component distributor in UK
INFO: Under construction
EMAIL: sales@golledge.co.uk
HTTP://www.golledge.co.uk/
RATING: (4)

Gowanda Electronics Corp.
Inductive components manufacturer
INFO: Products summary, contact
EMAIL: sales@gowanda.com
HTTP://www.gowanda.com/
RATING: (7)

GP Batteries (US) Inc.
Specialty battery supplier
INFO: Product info
EMAIL: info@gpbatteries.com
HTTP://www.gpbatteries.com/
RATING: (7)

Graychip
DSP chips and systems
INFO: Company/ product info, apps, datasheets
EMAIL: sales@graychip.com
HTTP://www.graychip.com
RATING: (9)

Gray's Semiconductor Page
From the personal Website of Gray Creager
INFO: URLs and info for semiconductor companies and the ICs they make
EMAIL: gcreager@scruznet.com
HTTP://www.scruznet.com/
~gcreager/hello5.htm
RATING: (9)

Green CirKit Homepage
A resource for PCBs, instruments, etc.

INFO: Source, resources, homespun
EMAIL: tntsupp@thinktink.com
HTTP://www.thinktink.com/
RATING: (7)

Greenweld Electronic Components
Surplus for hobbyists in UK
INFO: Inventory, links, contact
EMAIL:
100014.1463@compuserve.com
HTTP://www.herald.co.uk/clients/
G/Greenweld/
RATING: (7)

Greyden Press
Demand digital printing company
INFO: Quote request, print on demand info
EMAIL: benno@greyden.com
HTTP://www.greyden.com
RATING: (9)

GTC Sales
Semiconductor distributor
INFO: Linecard, inventory, locations
EMAIL: gtc@primenet.com
HTTP://www.gtcsales.com/
RATING: (7)

Guru Technologies, Inc.
Electronics design consulting
INFO: Clients, services info, jobs
EMAIL: mktg@gurutech.com
HTTP://www.gurutech.com/
RATING: (7)

G - H

The Guru's Lair
Don Lancaster's homepage
INFO: Don is an electronics columnist/author. Links, book info, reprints, you name it! A true information page.
EMAIL: don@tinaja.com
HTTP://www.tinaja.com/
RATING: (10)

H C Protek
Electronics instruments
INFO: Product info, DMM software, dealers list
EMAIL: hcprotek@aol.com
HTTP://www.techexpo.com/WWW/hcprotek
RATING: (9)

H R Diemen
Split diode transformers
INFO: Cross-ref to their parts
EMAIL: HR@hrdiemen.es
HTTP://www.hrdiemen.es/
RATING: (7)

Halfin — Belgium
Distributor of semiconductors/electron tubes
INFO: Stock, discontinued parts, order info, etc.
EMAIL: halfin@net-shopping.com
HTTP://www.wp.com/halfin/
RATING: (7)

Ham Radio Column in Popular Electronics
Joseph J. Carr
EMAIL: carrjj@aol.com

Ham Radio FTP
HTTP://oak.oakland.edu/pub/misc/hamradio/

Hamilton Hallmark
SEE ALSO: *Avnet*
EE resources
INFO: Products, support, tech
EMAIL: webmaster@tsc.hh.avnet.com
HTTP://www.tsc.hh.avnet.com/
RATING: (9)

Hammond Electronics and Subsidiaries
MRO & OEM distributor, assembly solutions, etc.
INFO: Locations, overview, products
EMAIL: webmaster@hammondelec.com
HTTP://www.hammondelec.com/
HTTP://www.hammondelec.com/emspage.html
RATING: (8)

Hantronix
LCD display manufacturer
INFO: Datasheets, apps, tech talk, reps
EMAIL: info@hantronix.com
HTTP://www.hantronix.com/
RATING: (9)

Harris Semiconductor Corp.
Semiconductor manufacturer
INFO: support, tons of product apps/info/datasheets
EMAIL: centapp@harris.com — for sales, see site
HTTP://www.semi.harris.com
RATING: (10) *****

Harry Krantz, Inc.
Components distributor
INFO: Part search & inventory, quotes
EMAIL: sales.hkc@mail.HarryKrantz.com
HTTP://www.harrykrantz.com/
FTP://ftp.harrykrantz.com/pub/
RATING: (9)

Hawk Electronics
Value-added distributor
INFO: Linecard, specials
EMAIL: sales@hawkusa.com
HTTP://www.hawkusa.com/
RATING: (7)

Hennessy Products
Enclosures manufacturer
INFO: Product info with pics, custom
EMAIL: See site for form
HTTP://www.hennessyproducts.com/
RATING: (7)

Hewlett-Packard Company
Computer, electronics testing products/components manufacturer
INFO: Top-level company site full of info
EMAIL: See site for form
HTTP://www.hp.com/
RATING: (10)*****

Hi-Tech Software
Embedded software tools
INFO: Tech docs, demos, free software
EMAIL: hitech@hitech.com.au
HTTP://www.hitech.com.au
FTP://ftp.hitech.com.au
LISTSERV: htc-request@hitech.com.au
RATING: (8)

High-Tech Direct
Direct marketer of components and products
INFO: Linecard, products carried
EMAIL: info@htd.com
HTTP://www.htd.com/
RATING: (7)

Hitachi America, Ltd.
Computers, semiconductors, instruments manufacturer
INFO: Product info, research, you name it!
EMAIL: See site for info
HTTP://www.halsp.hitachi.com/
HTTP://www.hitachi.co.jp/
RATING: (10)

Hitex Development Tools
Emulators, simulators, etc.
INFO: Very comprehensive uC/uProcessor page with links to all the resources
EMAIL: info@hitex.com
HTTP://www.hitex.com/
FTP://ftp.hitex.com/
RATING: (10) *****

HLK & Associates, Inc.
Brokers for electronics parts
INFO: Linecard, contact
EMAIL: sales@hlkassoc.com
HTTP://www.hlkassoc.com/
RATING: (7)

HLK Industrial, Inc.
Electronics distributor
INFO: Specials, products listing, links
EMAIL: sales@hlkind.com
HTTP://www.hlkind.com/
RATING: (8)

H

Home Automation Systems, Inc.
Home automation products
INFO: Online product info, news, dealers
EMAIL: custsvc@smarthome.com
HTTP://www.techmall.com/
smarthome
RATING: (10)

Home Automator Newsletter
By Jeff Anderson
INFO: Home automation information, articles, interviews, etc. Great site!
EMAIL: jeff@automator.com
HTTP://www.automator.com/
RATING: (10)

Home Controls, Inc.
EMAIL: homectrl@electriciti.com

Home Page Automation and Process Control
Information site for automation process control
INFO: Loaded with automation info, FAQs, links, etc.
EMAIL: wwwadmin@www.ba-karlsruhe.de
HTTP://www.ba-karlsruhe.de/
automation
HTTP://www.hitex.com/automation
RATING: (10)

Home Power Magazine
Online companion to magazine
INFO: Classifieds, links, covers, you name it!
EMAIL: See site for listing
HTTP://www.homepower.com/
RATING: (9)

Honeywell
Sensing & control
INFO: Product info/news/apps
EMAIL: info@micro.honeywell.com
HTTP://www.homecom.com/
sensing/sensing.html
RATING: (8)

How to Control HD44780-Based Character-LCD
Peer Ouwehand homepage
INFO: Extensive data on LCD display interfacing
EMAIL: pouweha@iaehv.nl
HTTP://www.iaehv.nl/users/
pouweha/f-lcd.htm
RATING: (9)

Howard W. Sams & Company
Publishers of electronics books, PHOTOFACT schematics, catalogs, etc.
Includes PROMPT Publications
Previously owned by Macmillan
Publishing — Now owned by Bell Atlantic
INFO: Complete site for book titles and news. Includes Dataview™ — The online catalog of the future!
EMAIL: hwsams@badg.com
HTTP://www.hwsams.com/
RATING: (As this company prints this book, I will leave this rating to you.)

HTH Online Catalog
Distributing electronic components
INFO: Tons of electronics info and hackers tricks
EMAIL: info@hth.com
HTTP://www.hth.com/
RATING: (9)

Hubbell, Inc.
Wiring devices manufacturer
INFO: FAQ, products, catalog, distributors
EMAIL: See site for listing
HTTP://www.hubbell-wiring.com/
RATING: (10)

Huber+Suhner, Inc.
RF & microwave products manufacturer
INFO: Jobs, order catalog
EMAIL: info@hubersuhner.com
HTTP://www.hubersuhner.com
RATING: (3)

Hughes Electronics Corporation
Aerospace, automotive, telecommunications products
INFO: R&D, news, links to subsidiaries
EMAIL: webmastere@hughes.com
HTTP://www.hughes.com/
RATING: (9)

Hughes Research Laboratories
Hughes Electronics research labs
INFO: Articles, technical labs, jobs. Great site to check the latest electronics technology!
EMAIL: webmaster@hrl.com
HTTP://www.hrl.com/
RATING: (10)

Huntsville Microsystems, Inc.
In-circuit emulators
INFO: Product info/datasheets/ drawings/demo
EMAIL: sales@hmi.com
HTTP://www.hmi.com/
FTP://ftp.hmi.com
RATING: (9)

Hutch & Son Industrial Electronics
Electronics distributor
INFO: Contact, linecard, links
EMAIL: ttaylor@hutch-and-son.com
HTTP://www.hutch-and-son.com/
RATING: (6)

Hybricon Corp.
Enclosures manufacturer
INFO: Product info, custom services offered
EMAIL: info@hybricon.com
HTTP://www.hybricon.com/
RATING: (8)

I-Cube, Inc.
Semiconductor switching solutions
INFO: FAQs, apps, product data
EMAIL: marketing@icube.com
HTTP://www.icube.com/
FTP://ftp.icube.com/pub/icubepdf/
BBS: 408-986-1652
RATING: (10)

I-Tech Corp.
Test equipment for peripherals
INFO: Product info, reps, articles
EMAIL: See site for form
HTTP://www.i-tech.com/
RATING: (7)

The I.C. Shop, Inc.
IC mask design
INFO: Services info, links
EMAIL: icshop@cyborganic.net
HTTP://www.cyborganic.com/people/icshop
RATING: (7)

I

IAC Industries
Manufacturer of workbenches, etc.
INFO: Product info with dimensions/pics
EMAIL: iacind@earthlink.net or iac@torrance.usa.com
HTTP://www.marketlink.com/iac/
RATING: (7)

IBM Corp.
Computer and business products, etc — Manufacturer
News, products, support, network, industry solutions. Huge information/software filled site.
EMAIL: askibm@www.ibm.com
HTTP://www.ibm.com/
RATING: (10)

IBM Corporation's Microelectronics Division
Semiconductor manufacturing
INFO: Full information of products/services
EMAIL: askibm@www.ibm.com
HTTP://www.chips.ibm.com/
RATING: (10)

IBS Electronics, Inc.
Components, value-added service, consumer
INFO: Linecards, inventory, specials
EMAIL: ibs@ibs-elec.com
HTTP://www.ibs-elec.com/
RATING: (8)

ICE Technology, Inc.
Manufactures chip programmers/emmulators
INFO: Product info, demo software, support
EMAIL: webmaster@icetech.com
HTTP://www.icetech.com/
FTP:// icetech:icetech@ftp.icetech.com/pub/
RATING: (9)

Idaho National Engineering Lab (INEL)
Department of Energy engineering lab
INFO: News, whitepages, programs, fun
EMAIL: See whitepages on site
HTTP://www.inel.gov/
RATING: (8)

Ideal Industries, Inc.
Wire, products, tools
HTTP://www.industry.net/ideal.industries

IDEC
HTTP://www.industry.net/ideccorp/

IEE — Institute of Electrical Engineers
Largest professional engineering society in Europe
INFO: Membership, library, links, archives. Just about everything!
EMAIL: postmaster@iee.org.uk
HTTP://www.iee.org.uk/
RATING: (10) *****

IEEE, Inc.
The Institute of Electrical and Electronics Engineers, Inc.
INFO: Membership, FAQ, full site search. The Engineering site!
EMAIL: See site for forms
HTTP://www.ieee.org/
FTP://ftp.ieee.org

GOHPER://gopher.ieee.org
SEE ALSO: Directory of
Organizations for further IEEE
addresses
RATING: (10)

IEEE Webpage - University of Michigan
Student branch
INFO: Chapter info, links,
schedule, etc.
EMAIL: See site for form
HTTP://www.engin.umich.edu/soc/
ieee/
RATING: (7)

Imagenation Corporation
Frame grabber cards
INFO: Software, apps, product info
EMAIL: info@imagenation.com
HTTP://www.imagenation.com/
RATING: (8)

Images Co.
Hobby/robot information, projects,
supplies
INFO: Projects, catalog request
EMAIL: imagesco@he.net
HTTP://www.imagesco.com
RATING: (7)

Imminent Technologies
Distributor of semiconductors and
other parts
INFO: Linecard, links
EMAIL: rbiti@earthlink.net
HTTP://home.earthlink.net/~rbiti/
RATING: (6)

IMP, Inc. - International Microelectronic Products
Manufactures analog signal
processing ICs. Invented EPAC
INFO: EPAC data, software, apps,
news, company info

EMAIL: info@impweb.com
HTTP://www.impweb.com
RATING: (10)

Impact Components
Independent electronics distributor
INFO: Helpful info for buyers
EMAIL:
impactinfo@impactcomponents.com
HTTP://
www.impactcomponents.com/
RATING: (9)

Industrial Battery Products, Inc.
INFO: Buying guide, battery info
EMAIL: ibpstl@mail.mo.net
HTTP://www.ibpstl.com/
RATING: (6)

Industrial Computer Source Book
HTTP://www.industry.net/
indcompsrc/

Industrial Electronics, Inc.
General-line distributor
INFO: Linecard, specials
EMAIL: iei@flash.net
HTTP://www.flash.net/~iei/
RATING: (7)

Industry.Net
ISP/Host for several technology
companies
INFO: Listings to companies, orgs
hosted. You are able to view some
information on the company and
Email or fax them online.
EMAIL: See site

I

HTTP://www.industry.net/
RATING: (see note)
NOTE: Companies listed on
Industry.Net in this book are not
rated.

Inform Software Corporation
Fuzzy logic design tools
INFO: FuzzyTECH info, apps,
resources. Great fuzzy resource
page worth the time to look all over.
EMAIL: fuzzy@informusa.com
HTTP://www.inform-ac.com
FTP://ftp.inform-ac.com
RATING: (10)

Information Storage Devices (ISD)
Digital recording ICs manufacturer
INFO: Product info/support/apps,
jobs
EMAIL: info@isd.com
HTTP://www.isd.com/
RATING: (10)

Information SuperLibrary
HTTP://www.mcp.com/

Information Unlimited
Experimental supplies distributor
INFO: Projects, books offered, order
EMAIL: wako2@xtdl.com
HTTP://www.amazing1.com/
RATING: (9)

Inland Electronic Suppliers
Distributor of electronics,
equipment
INFO: Products carried, locations
EMAIL: inland@inland-
electronics.com

HTTP://www.inland-
electronics.com/
RATING: (5)

Inland Empire Components, Inc.
Buy/sell components
INFO: Inventory search, links
EMAIL: inland@earthlink.net
HTTP://www.inlandempcomp.com/
RATING: (5)

Inotek Technologies Corp.
Process controls/instrumentation
distributor
INFO: Cyber-catalog, conversion
tables. Helpful site!
EMAIL: info@inotek.com
HTTP://www.inotek.com/
RATING: (9)

Inovonics, Inc.
Manufacturers of broadcast
components
INFO: Product info, distributors
list, prices
EMAIL: info@inovon.com
HTTP://www.inovon.com/
RATING: (7)

Insight Knowledge Networks
Distributor with 35 locations
INFO: Linecard, locations, links
EMAIL: See site form
HTTP://www.ikn.com
RATING: (9)

Institute for Signal and Information Processing
R&D for speech/pattern recog. &
DSP
INFO: Speech recognition, public
domain software

EMAIL:
webmaster@isip.msstate.edu
HTTP://www.isip.msstate.edu/
FTP://ftp.isip.mstate.edu/
RATING: (10)

Integrated Circuit Design Concepts

IC design assistance
INFO: Cell libraries, service info
EMAIL: icdcbob@netcom.com
HTTP://universe.digex.net/~icdc/
FTP://ftp.netcom.com/pub/ic/icdc/
RATING: (8)

Integrated Circuit Systems, Inc. (ICS)

Clock synthesizers, ASICs, etc.
INFO: Product info, datasheets
EMAIL: webmaster@icsinc.com
HTTP://www.icsinc.com/
RATING: (8)

Integrated Device Technology, Inc. (IDT)

Microprocessors, memory and logic
manufacturer
INFO: News, product library with
datasheets, apps
EMAIL: ussales@idt.com
HTTP://www.idt.com/
RATING: (9)

Integrated System Design

Online *EEDesign* magazine
INFO: Great articles, archives, ads
EMAIL: mitch@asic.com
HTTP://www.isdmag.com
RATING: (10)

Integrated Systems, Inc. (ISI)

Manufactures support software for
uCs
INFO: Support, papers, company/

product info, training/show
schedules
EMAIL: webmaster@isi.com
HTTP://www.isi.com
RATING: (9)

Integrated Telecom Technology, Inc. (IgT)

High-speed networking devices
manufacturer
INFO: Docs, company/product info
EMAIL: products@igt.com
HTTP://www.igt.com/
RATING: (9)

Intel Corp.

Microprocessor, microcontroller
and other semiconductors
manufacturer
INFO: Just about everything. Intel
wrote the book on a perfect
site for product support.
EMAIL: See site for listing
HTTP://www.intel.com/
FTP://ftp.intel.com/pub/
RATING: (10) *****

Intelligent Instruments

Supplier of data aquisition
software/hardware
INFO: Extensive product info/
support, demos
EMAIL:
webmanager@instrument.com
HTTP://www.instrument.com/
FTP://ftp.instrument.com/pub/
RATING: (10) *****

Intellon Corp.

BBS: 904-237-8841

I

Interactive CAD Systems
EDA, CAD software
INFO: Product info, demos, FAQs
EMAIL: sales@icadsys.com
HTTP://www.icadsys.com/
RATING: (7)

Interactive Filter Design
Page by Tony Fisher
INFO: A great web utility to design filters
EMAIL: fisher@minster.york.ac.uk
HTTP:/dcpu1.cs.york.ac.uk:6666/fisher/mkfilter/
RATING: (9)

Interactive Image Technologies, Ltd.
SEE: *Electronics Workbench*

Intercomp, Inc.
Components distributor for OEM/distributors
INFO: Components, inventory
EMAIL: intrcomp@icanect.net
HTTP://www.icanect.net/intcmp/
RATING: (6)

Intergraph Corporation
Personal workstations, PCs, technical software
INFO: Products, user groups, demos, international info
EMAIL: webmaster@intergraph.com
HTTP://www.intergraph.com/
RATING: (10)

Interlink Electronics
Mouses, remotes, other input technology
INFO: News, product info with pics, jobs
EMAIL: support@interlinkelec.com
HTTP://www.interlinkelec.com/
RATING: (8)

Intermec Corp.
Barcoding solutions manufacturer
INFO: Company/ product info
EMAIL: support@intermec.com
BBS: 1-206-356-1811
HTTP://www.intermec.com/
RATING: (7)

International Circuit Sales (ICS)
Franchised distributor
INFO: Linecard, contact
EMAIL: ics@icsstock.com
HTTP://www.icsstock.com/
RATING: (8)

International Power Devices, Inc. (IPD)
EMAIL: ipsales@ipd-hq.ccmail.compuserve.com

International Rectifier Corp.
Power semiconductors
INFO: Tech apps, sheets, tips, company/product info
EMAIL: irf-webmaster@irf.com or see site
HTTP://www.irf.com/
RATING: (9) ***

International Semiconductor, Inc.
Manufactures discrete components
INFO: Company site
EMAIL: ISISEMI@prodigy.com
HTTP://www.multiplex.com/1/a/intlsemiconductor/
RATING: (3)

International Society of Certified Electronic Technicians (ISCET)
SEE ALSO: NESDA
INFO: Organization site, CET certification, etc.
EMAIL: iscetfw@nesda.com
HTTP://www.iscet.org
RATING: (n/a)

Internet Reference — The English Server
INFO: Links to tons of Internet resources such as libraries, patent, etc.
HTTP://english-www.hss.cmu.edu/reference/
SEE ALSO: http://english-www.hss.cmu.edu/
RATING: (10)

Interpoint Corp.
EMAIL: power@intp.com

Intusoft
EDA software tools (simulation, filter/magnetic synthesis, test software design)
INFO: SPICE models, evaluation software, newsletter, demos, articles, apps, notes
EMAIL: info@intusoft.com
HTTP://www.intusoft.com/
FTP://ftp.intusoft.com
RATING: (9)

Iomega Corp.
Zip drives
INFO: Newest software and info
EMAIL: support@iomega.com
HTTP://www.iomega.com/
RATING: (9)

IPC — Institute for Interconnecting and Packaging Elecronics Circuits
SEE: *Automata, Inc.*
HTTP://www.automata.com/ipc/

IRC Tresco
Master distributor of passives
INFO: Product lines, contact
EMAIL: tresco@tresco.com
HTTP://www.tresco.com/
RATING: (5)

Ironwood Electronics
VLSI interconnection specialists
INFO: Products, online ordering, distributor list
EMAIL: See site for form
HTTP://www.tc.frontiercomm.net/~mannan/
RATING: (7)

Irvine Sensors Corp.
Sensors/ICs manufacturer
INFO: Product info, news, company
EMAIL: lomara@irvine-sensors.com
HTTP://www.irvine-sensors.com/
RATING: (9)

ISO — Insurance Services Office, Inc.
Supplier of statistical, actuarial, and underwriting information
INFO: Full ISO info, links, services, studies
EMAIL: See site for form
HTTP://www.iso.com/
RATING: (9)

ITT Cannon
Manufactures connectors
INFO: Product info/support/tips
EMAIL: webmaster@ittcannon.com

I - J

HTTP://www.ittcannon.com/
RATING: (8)

ITT Semiconductors
DSP, and other complex ICs
INFO: Product info/datasheets, jobs
EMAIL: info@itt-sc.de
HTTP://www.itt-sc.de/
RATING: (7)

ITU Technologies
MCU products for hobby/
developement
INFO: Specials, product/company
info, software, support, mega-links
EMAIL: sales@itutech.com
HTTP://www.itutech.com/
RATING: (10)***

Ivex Design International
PCB design, layout tools
INFO: Software, support, tech
forum, links, etc.
EMAIL: info@ivex.com
HTTP://www.ivex.com/
RATING: (9) ***

IXYS Corp.
Power semiconductor manufacturer
INFO: Support, product info
EMAIL: ixyscorp@aol.com
HTTP://www.ixys.com/
RATING: (7)

J.R. Kerr Prototyping Development
INFO: PIC-SERVO DC servo motor
control
EMAIL: dcservo@aol.com
HTTP://users.aol.com/dcservo/
consult.html
RATING: (7)

Jaco Electronics
Worldwide distributor of
components and computer
peripherals
INFO: Linecard, news, extensive
development support
EMAIL: jaco@webscope.com
HTTP://www.jacoelectronics.com/
RATING: (9) ***

Jamark International
Components distributor
INFO: Locations, linecard
EMAIL: east@jamark.com,
west@jamark.com
HTTP://www.jamark.com/
RATING: (6)

Jameco Electronics
Catalog components distributor
INFO: Linecard, new, specials, jobs
EMAIL: info@jameco.com
HTTP://www.jameco.com/
RATING: (7)

Jason McDonald
SEE: *EG3*
EMAIL: jason@eg3.com

Javanco
Mail order surplus components
INFO: Product line, inventory, links
EMAIL: javanco@javanco.com
HTTP://www.javanco.com/
RATING: (9)

JDR Microdevices
Mail order components &
equipment
INFO: Online shopping with cart,
specials, catalogs to contests,
articles, fun. This site is TRUE
shopping on the Internet. Awesome
concept!

EMAIL: webideas@jdr.com
HTTP://www.jdr.com/
RATING: (10) **********

JECH TECH, Inc.
Manufactures telephone electronics
INFO: Product info/specs, help
pages
EMAIL: jectech@infinet.com
HTTP://www.infinet.com/~jectech
RATING: (8)

Jerome Industries Corp.
EMAIL: jeromeind@aol.com

Jewell Electrical Instruments
Sensors, panel meters, avionics
manufacturer
INFO: New, product info, specs
EMAIL: sales@teamjewell.com
HTTP://www.teamjewell.com/
RATING: (7)

Jim's Crosley Antique Radio Page
INFO: Crosley Radios pics/model
numbers, etc. Great site for
antique radio pics
EMAIL: oldradio@gulf.net
HTTP://www.pcola.gulf.net/
~oldradio/
RATING: (10)

JKL Components Corp.
Manufactures mini-lightbulbs/
flourescents/LEDs
INFO: Product pics, contact
EMAIL: jkl@jkllamps.com
HTTP://www.jkllamps.com/
RATING: (6)

JME Electronics
Independant distributor
INFO: Some stock info

EMAIL: n/a
http://www.drive.net/biz/jme.htm
RATING: (3)

JMI Software Systems, INC.
ROMable kernels for embedded
systems
INFO: Contacts, product info
EMAIL: sales@jmi.com
HTTP://www.mcb.net/jmi/
RATING: (7)

John Wiley & Sons, Inc.
Book/electronic media publisher
INFO: Online products/services,
authors info. Great site for journals/
books, etc.
EMAIL: info@qm.jwiley.com
HTTP://wiley1000.wiley.com/
FTP://ftp.wiley.com/pub/
RATING: (10)

Johnnie Walker's Home Page
INFO: PIC hardware, software,
links, projects
EMAIL: ceejbcw@cee.hw.ac.uk
HTTP://www.cee.hw.ac.uk/
~ceejbcw/
RATING: (9)

Johnson Controls Battery Group
Lead-acid sealed
INFO: Tips, products, new
technology, locations
EMAIL: bg.webmaster@jci.com
HTTP://198.81.196.4/bg/
default.htm
RATING: (9)

J - K

Journal of Artificial Intelligence Research
JAIR is a refereed journal for scientific papers
INFO: Articles, FAQ, links, publish
EMAIL: jair-editor@ptolemy.arc.nasa.gov
HTTP://www.cs.washington.edu/research/jair/home.html
RATING: (10)

Journal of Computer Controlled Systems
PC/104 and misc. buses information magazine
INFO: Database, links, papers, you name it!
EMAIL: editor@controlled.com
HTTP://www.controlled.com/
RATING: (10) ***

JTM Enterprises
IC & semiconductor distributor
INFO: Linecard, inventory, links
EMAIL: jtm@mail.jtment.com
HTTP://www.jtment.com/
RATING: (8)

Judd Wire, Inc.
INFO: News, product info/specs
EMAIL: sales@juddwire.com
HTTP://www.juddwire.com/
RATING: (7)

Just In Time ICs
Distributor of ICs, passives, etc.
INFO: Products and prices, contact
EMAIL: justintime@batnet.com
HTTP://www.batnet.com/justintime/JUTIC.html
RATING: (7)

Kaman Instrumentation Corp.
Position sensor manufacturer
INFO: Sensor info, apps, datasheets
EMAIL: info-cos2@kaman.com
HTTP://www.rmi.net/kaman/
RATING: (9)

Kangaroo Tabor Radio
Amateur radio software
INFO: Software info, links
EMAIL: ku5s@wtrt.net
HTTP://ourworld.compuserve.com/homepages/KU5S
HTTP://www.wtrt.net/~ku5s/
RATING: (7)

Keil Software
8051, 251, 166 compilers
INFO: Demos, FAQs, tech, links
EMAIL: sales@keil.com
HTTP://www.keil.com/
FTP://ftp.keil.com
RATING: (10)

Keithley Instruments, Inc.
Precision instruments & systems manufacturer
INFO: Specs, info, newsletter
EMAIL: product_info@keithley.com
HTTP://www.keithley.com/
FTP://ftp.keithley.com/pub/instr/
RATING: (10)

KEMA Information Center
ISO qualifications help
INFO: Complete info/help on ISO
EMAIL: info@kema.be
HTTP://www.kema.be/kema/default.nclk
RATING: (10)

Kemet
Capacitors manufacturer
INFO: Product info, FAQ, service

EMAIL: capmaster@kemet.com
HTTP://www.kemet.com/
RATING: (7)

Kent Electronics
National distributor
INFO: Linecard, library, locations
EMAIL: sales@kentelec.com
HTTP://www.kentelec.com/
RATING: (9)

Kenwood Communications Corp.
Amateur radio, land mobile radio manufacturer
INFO: Product info, news, dealers
EMAIL: paulem@ix.netcom.com
HTTP://www.kenwood.net
RATING: (8)

Kepco, Inc.
Electronically-controlled power supplies
INFO: Software for plug&play, drivers, product info/handbooks
EMAIL: hq@kepcopower.com
HTTP://www.kepcopower.com/
RATING: (9)

Kester Solder
Solder and solder products manufacturer
INFO: Products/services info, datasheets
EMAIL: webmaster@kester.com
HTTP://www.kester.com/
RATING: (9)

Keystone Electronics Corp.
Interconnectors and PCB hardware
INFO: News, products, distributor list
EMAIL: contactus@keyelco.com
HTTP://www.keyelco.com/
RATING: (9)

KineticSystems Corp.
Manufacturer of data acquisition hardware/software
INFO: Datasheets/notes/apps/ software
EMAIL: mkt-info@kscorp.com
HTTP://www.kscorp.com
RATING: (8)

Kingbright USA Corp.
LED manufacturer
INFO: News, reps, tech info
EMAIL: info@Kingbright-led.com
HTTP://www.kingbright-led.com/
RATING: (6)

Kistler Instruments Corp.
Piezoelectric-based sensors, etc.
INFO: Product info, contact, theory
EMAIL: webmaster@kistler.com
HTTP://www.kistler.com/
RATING: (7)

Kluwer Academic Publishers
Publisher in the Netherlands
INFO: Journals catalog, machine learning online
EMAIL: services@wkap.nl
HTTP://www.wkap.nl/
RATING: (7)

Kolibri Industries
PCB and design service
INFO: under construction
EMAIL: kolibri@portal.ca
HTTP://www.portal.ca/~kolibri/
RATING: (4)

KV Electronics
Electronics distributor
INFO: Linecard, contact

K - L

EMAIL:
kvsales@caledonia.polaristel.net
HTTP://www.electronet.com/
kvhome.htm
RATING: (3)

Kycon Cable & Connector, Inc.
IC sockets, cables, etc.
INFO: Full catalog with specs and pics
EMAIL: sales@kycon.com
HTTP://www.kycon.com/kycon
RATING: (10)

L J Enterprises
Distributor of ICs, diodes, transistors
INFO: Linecard, small stock list
EMAIL: lee@ljent.com
HTTP://www.ljent.com/
RATING: (5)

L.O.S.A
List of Stamp Applications
INFO: Basic Stamp stories, help
EMAIL: cj@hth.com
HTTP://www.hth.com/losa.htm
RATING: (7)

Lake DSP Pty. Ltd.
Advanced digital audio using DSPs
INFO: Datasheets, products, specs, links, articles, demos, DSP paradise!
EMAIL: info@lakedsp.com.au
HTTP://www.lakedsp.com/
index.htm
RATING: (10) ***

Lakeland Batteries, Inc.
Battery distributor/exporter
INFO: Products carries, links,
EMAIL: lakeland@battery-usa.com
HTTP://www.battery-usa.com/
RATING: (7)

Lama Electronics, Inc.
Cables, mobile radios, power supplies
INFO: Products carried, company info
EMAIL: lama@ix.netcom.com
HTTP://lamaelectronics.com/
RATING: (7)

Lamp Technology
Bulbs galore!
INFO: Linecard, online ordering, pricing
EMAIL: lamptech@webscope.com
HTTP://www.webscope.com/
lamptech/info.html
RATING: (7)

Landmark Electronics Corp.
Electronics & equipment distributor
INFO: Company and line info
EMAIL:
michael.bridge@sbaonline.gov
HTTP://www.inc.com/users/
mbridge1.html
RATING: (4)

Lansdale Semiconductor, Inc.
Integrated circuit manufacturer
INFO: Company/product info, news, jobs
EMAIL: lansdale@lansdale.com
HTTP://lansdale.com/~lansdale
RATING: (9)

Lashen Electronics, Inc.
Distributor of parts, supplies, equipment

INFO: Line Card products, specials & info
EMAIL: sales@lashen.com
HTTP://www.lashen.com/
RATING: (7)

Lattice Semiconductor Corp.
PLDs and related software
INFO: Apps, products, datasheets, files, literature
EMAIL: See site for listing
HTTP://www.latticesemi.com/
FTP: access through http://www.latticesemi.com/ftp/
RATING: (10)

Lawrence Livermore National Laboratory
University of California — Department of Energy
INFO: Documents, search, projects, publications, links
EMAIL: See site for departments
HTTP://www.llnl.gov/
RATING: (10) *****

LDG Electronics
68HC11 dev. sys. and amateur radio-based products
INFO: Product info, 68HC11 data
EMAIL: ldg@radix.net
HTTP://www.radix.net/~ldg/
RATING: (7)

Leader Electronics, Inc.
General-line distributor
INFO: Linecard, company info
EMAIL: see site for listing
HTTP://www.leaderelectronics.com/
RATING: (7)

LeCroy Corp.
Instruments and DSO manufacturer
INFO: FAQs, tutorials, specs/apps

EMAIL: webmaster@lecroy.com
HTTP://www.lecroy.com/
RATING: (9)

LEDvision, Inc.
Color LED scrolling and graphic signs
EMAIL: sales@capital-elec. com
led@capital-elec.com for electronics brochure
HTTP://www.capital-elec.com
RATING: (9)

Leff Electronics
Full-line parts distributor and EDGE member.
INFO: Links to line, contact
EMAIL: leff@nb.net
HTTP://www.leff.com/
RATING: (6)

Leo Electronics, Inc.
Technology trading & manufacturing company
INFO: Linecard, inventory, specials
EMAIL: dbecker@excess.com
HTTP://www.excess.com/
RATING: (9)

Level One Communications
Mixed signal ICs
INFO: Product info/apps, sales offices
EMAIL: webmaster@level1.com
HTTP://www.level1.com/
RATING: (6)

Lighting Specialties Company
Distributor of illuminated magnifiers
INFO: Product info with pics

L

EMAIL: lightspc@interaccess.com
HTTP://www.lightspc.com/
RATING: (8)

Linear Electronics Troubleshooter

Electronics projects page by Clayton Forrester
INFO: Linear electronic projects with PCB art and all! New projects often.
EMAIL: n/a
HTTP://www.geocities.com/ CapeCanaveral/5322/
RATING: (9)

Linear Technology Corp.

Semiconductor manufacturer
INFO: Under construction as of Sept. 1, 1996
EMAIL: n/a
HTTP://www.linear-tech.com/
RATING: (n/a)

Link Technologies, Ltd.

Canadian electronics manufacturer
INFO: Product and service info
EMAIL: info@linktech.ca
HTTP://www.linktech.ca/
RATING: (7)

Lipidex Corp.

Embedded controller-based product design
INFO: PIC Shareware, sample products designed by Lipidex
EMAIL: jason@ultranet.com
HTTP://www.ultranet.com/~jason
RATING: (9)

List of USENET FAQs

OSU sponsered
INFO: Listing of FAQs on USENET including electronics subjects. Excellent reference to bookmark.
EMAIL: webmaster@cis.ohio-state.edu
HTTP://www.cis.ohio-state.edu/ hypertext/faq/usenet/top.html
RATING: (9)

Littelfuse

HTTP://www.industry.net/ littelfuse.electronics/

Logic Devices, Inc.

High-performance digital ICs manufacturer
INFO: Product info/support/apps/ datasheets, jobs, etc.
EMAIL: See site for listing
HTTP://www.logicdevices.com/
RATING: (9)

Logical Devices, Inc.

Device programmers manufacturer
INFO: Product info, software
EMAIL: logdev@henge.com
HTTP://www.logicaldevices.com/
FTP://ftp.gate.net/pub/users/logdev/
RATING: (9)

Low Voltage Home Pre-Wire Guide

Mark Henrichs' page
INFO: Low voltage installation primer
EMAIL: markh@mcdata.com
HTTP://www.mcdata.com/ ~meh0045/homewire/ wire_guide.html
RATING: (10)

Lowe Electronics Ltd.
Ham radio/scanners/shortwave
INFO: Equipment reviews/user
notes
EMAIL: info@lowe.co.uk
HTTP://www.lowe.co.uk/
RATING: (8)

LPB, Inc.
Broadcast equipment manufacturer
INFO: Contact, pics
EMAIL: lpbsales@lpbinc.com
HTTP://www.lpbinc.com
RATING: (3)

LPS Labs
HTTP://www.industry.net/lps.labs

LSI Logic Corp.
System-on-a-chip custom
semiconductors
INFO: Online database, jobs,
product info
EMAIL: See site for directory and
form
HTTP://www.lsilogic.com/
RATING: (10)

Lucent Technologies
AT&T's systems and technology
company
INFO: Descriptions of projects,
news, journals, magazines, you
name it!
EMAIL: webmaster@att.com
HTTP://www.att.com/lucent/
RATING: (10)

Luke Systems International
Distributor of obsolete AMD chips
INFO: New items, contact
EMAIL: parts@lukechips.com
HTTP://www.lukechips.com/
RATING: (7)

Lumex Opto/Components, Inc.
Opto and photo products
manufacturer
INFO: Catalog, tech notes. Full of
info!
EMAIL: sales@lumex.com
HTTP://www.lumex.com/
RATING: (9)

Lundahl Instruments
Ultrasonic sensors manufacturer
INFO: Apps, specs
EMAIL: lundahl@cache.net
HTTP://www.cache.net/lundahl/
RATING: (7)

Lutze Network
Wire/cable, power supplies, etc.,
manufacturer
INFO: Product info, downloads,
contest
EMAIL: lutzeinc@webserve.net
HTTP://www.lutze.com/
RATING: (8)

M/A-Com Phi
RF power transistors/modules
manufacturer
INFO: Product datasheets/apps,
jobs
EMAIL: macom-phi@clubnet.net
HTTP://www.macom-phi.com/
RATING: (7)

M & T Systems
Mobile robots
INFO: Product info, robotic links
EMAIL: mandtsys@ix.netcom.com
HTTP://www.netcom.com/
~mandtsys/robots.html
RATING: (7)

M

M.G. Electronics
Componets broker
INFO: Inventory, vendor application
EMAIL: mgelec@adnetsol.com
HTTP://secure.adnetsol.com/newmg/
RATING: (7)

Macmillan Computer Publishing
Electronic/Computer software & hardware publishing, including books, disks, etc.
Imprints include:
Que,
Sams,
Sams.Net,
New Riders,
Hayden,
Borland Press,
BradyGames,
Ziff-Davis,
Waite Group
INFO: Catalog, news, upcoming titles, etc.
EMAIL: Varies by region. Consult Web site.
HTTP://www.mcp.com/mcp/
RATING: (8)

Magnecraft & Struthers-Dunn
Manufactures relays
INFO: Catalog, news, tech, reps
EMAIL: info@magnecraft.com
HTTP://www.magnecraft.com/
RATING: (9)

MagneTek, Inc.
Electronic and electrical products manufacturer

INFO: Product info/tech notes, etc. Cute site!
EMAIL: productinfo@magnetek.com
HTTP://www.magnetek.com/
RATING: (9)

Major League Electronics
Connector manufacturer
INFO: Online catalog, distributor list
EMAIL: mle@iglou.com
HTTP://www.iglou.com/major_league/
RATING: (10)

Manhattan/CDT
Wire/cable manufacturer
INFO: Products, locations
EMAIL: See site
HTTP://www.manhattancdt.com/
RATING: (7)

Mar Vac Electronics
Distributor of parts/equipment
INFO: Specials, locations, linecard
EMAIL: sales@marvac.com
HTTP://www.marvac.com/
RATING: (7)

Marathon Special Products
HTTP://www.industry.net/marathon.special.products

Marquardt Switches, Inc.
Switch manufacturer
INFO: Catalog, product info/specs
EMAIL: info@marqswitch.com
HTTP://www.switches.com/
RATING: (7)

Marsh Electronics, Inc.
Broadline electronics distributor
INFO: Semiconductor/passive news, linecard

EMAIL: See site for listing
HTTP://www.execpc.com/marsh/
RATING: (7)

Marshall Industries
Electronics distributor
INFO: News, online shopping/
seminars, product info/catalog, tech
publications/software, it is ALL
HERE!
EMAIL://See site for form
HTTP://www.marshall.com/
RATING: (10) **********

MATLAB Based Books
Broadline technology books
INFO: Book listings
EMAIL: info@mathworks.com
HTTP://www.mathworks.com/
books/intro_books.html
RATING: (7)

Matsushita Electric Works R&D Labs, Inc.
Multiple company listings on site
INFO: Company sites
EMAIL: See sites for directory
HTTP://www.mew.com/
RATING: (6)

Matt Kellners Homepage & Dorsola's Webpage Works — DWW
INFO: Featuring the Tech Support
FAQ, a database of frequently-asked
questions about computers and their
related problems.
EMAIL:
mkellner@hertz.elee.calpoly.edu
HTTP://www.calpoly.edu/~mkellner
RATING: (9)

Matthew McDonald's Homepage
INFO: AI, robotic, PIC links

EMAIL: n/a
HTTP://www.cs.uwa.edu.au/
~mafm/robot/pic.html
RATING: (7)

Max Froding
RF filter design software
INFO: Software download for free
EMAIL: maxf@ll.mit.edu
HTTP://members.aol.com/maxfro/
private/
RATING: (9)

Maxim Integrated Products
Semiconductor manufacturer
INFO: New products, samples,
spice models, datasheets
EMAIL: webmaster@maxim-
ic.com
HTTP://www.maxim-ic.com/
RATING: (9)

Maxon Precision Motors
Motors and controller manufacturer
INFO: Great pics/data catalog,
locations
EMAIL: info@maxonmotor.com
HTTP://www.maxonmotor.com/
RATING: (9)

Maxtek Components Corp.
A Maxim/Tektronix company
INFO: Design, manufacture and
test of MCMs
EMAIL: technology@maxtek.com
HTTP://www.maxtek.com/
RATING: (7)

Maxvale International Ltd.
Hard-to-find products, JIT, kitting
INFO: Inventory, company info

M

EMAIL:
maxvale@maxvale.com.hk
HTTP://www.maxvale.com.hk/
RATING: (6)

McDonald Distributors
Electrical wholesale distributor
INFO: Links, linecard, local
distributors
EMAIL: mcdmail@mcdonald.com
HTTP://wahoo.netrunner.net/
mcdonald/
RATING: (8)

McGraw-Hill Companies
Publisher
INFO: McGraw's various subsidiary
sites, FAQ, directory, etc.
EMAIL: See site for directory
HTTP://www.mcgraw-hill.com/
RATING: (Over-all 10)

McMaster-Carr Supply Company
Publisher of a catalog containing
over 200,000 industrial items
INFO: sales offices, jobs
EMAIL:
webmaster@mcmaster.com
HTTP://www.mcmaster.com/
RATING: (5)

MCU/MPU FAQs and Overviews
SEE: *EG3* or *FAQ Library*

MECI
Surplus electronics
INFO: Online catalog/shopping
EMAIL: meci@meci.com
HTTP://www.meci.com/
RATING: (9)

Mentor Graphics Corp.
EDA/CAD software, etc.
INFO: Products, services, solutions,
support, successes
EMAIL: See site for forms
HTTP://www.mentorg.com/
RATING: (9)

Mentor Graphics — ePARTS
Storefront for EDA library products
INFO: News, products, contacts
EMAIL:
webmaster@eparts.mentorg.com
HTTP://eparts.mentorg.com/
HTTP://www.eparts.com/
RATING: (9)

Meredith Instruments
Laser components
INFO: Catalog, newletter, product
info
EMAIL: lasers1@ix.netcom.com
HTTP://www.mi-lasers.com/
RATING: (9)

Meridian Electronics
Stocking independent distributor
INFO: Linecard, contact
EMAIL: ic@meridianltd.com
HTTP://www.meridianltd.com/
RATING: (4)

Meritec
Division of Association Enterprises
76311.2313@compuserve.com

Meta-Software, Inc.
Software for IC design
INFO: News, products, university
programs, lit.
EMAIL: web@metasw.com
HTTP://www.metasw.com/
RATING: (8)

M

Metalinspec — Mexico
Distributes instruments/equipment
INFO: Linecard, product pics, links
EMAIL: metalins@intercable.net
HTTP://www.metalinspec.com.mx/
RATING: (7)

Metcal, Inc.
Soldering/Desoldering
manufacturer
INFO: Creator of SmartHeat
soldering and rework systems.
Training, support, software, custom
applications solutions, and fume
extraction — it's all here.
SEE ALSO: *The Rework & Repair
Exchange*
EMAIL: smartheat@metcal.com
HTTP://www.metcal.com/
home.html
RATING: (10)*****

Methode Electronics, Inc.
Manufactures component devices
INFO: Some product description
EMAIL: webmaster@methode.com
HTTP://www.methode.com/
RATING: (5)

Micatron
Independent stocking distributor
INFO: Specials, linecard
EMAIL: See site for request form
HTTP://www.micatron.com
RATING: (6)

Micro Care Corp.
Environmental progressive solvents
& tools
INFO: Links, guides, cost analysis
EMAIL: info@microcare.com
HTTP://www.microcare.com/
RATING: (10)

Micro Linear Corp.
EMAIL: info@ulinear.com

Micro Networks Clock
Manufactures data converters/amps/
oscil., etc.
INFO: Product info/apps/datasheets
EMAIL: sales@mnc.com
HTTP://www.mnc.com/
RATING: (8)

Microchip Technology, Inc.
Maker of the PIC microcontroller
INFO: A total information site with
links, datasheets, specs, you name
it!
EMAIL:
webmaster@microchip.com
HTTP://www.microchip.com/
RATING: (10)

MicroClock, Inc.
Clock chips manufacturer
INFO: Tech, apps, contact
EMAIL: sales@microclock.com
HTTP://www.microclock.com/
RATING: (5)

Microcode Engineering, Inc.
Free online demo of "CircuitMaker"
and "TraxMaker"
INFO: Demos, product info, FAQs,
orders
EMAIL:
74777.1144@compuserve.com
HTTP://www.microcode.com/
RATING: (8)

Microconsultants Home Page
Manufactures programmable
machine controllers & SPC tools

M

INFO: Product info, links, articles
EMAIL:
webmaster@microconsultants.com
HTTP://
www.microconsultants.com/
RATING: (8)

MICROCONTROLLER PRIMER FAQ
Huge FAQ by Russell Hersch
EMAIL: russ@shani.net or
russ@silicom.co.il
USENET: news.answers,
comp.arch.embedded
RATING FOR FAQ: (10)

Microcontroller FTP
INFO: Projects, assemblers,
downloads for 8051, PICs
FTP://ftp.mcc.ac.uk/pub/micro-
controllers/
RATING: (9)

Microelectronics Research Center, U of Idaho
University research page
INFO: VLSI, ASIC, FPGA, NASA
links
EMAIL: lharold@mrc.uidaho.edu
HTTP://www.mrc.uidaho.edu/
RATING: (8)

microEngineering Labs
PIC microcontroller development
tools
Site also contains AGCO, DENCO
SIGMA
INFO: PIC data and product info
EMAIL:
71165.322@compuserve.com
HTTP://www.melabs.com/
RATING: (7)

Micromint, Inc.
SEE ALSO: *Circuit Cellar
Magazine*
Embedded controller manufacturer
INFO: New products, datasheets,
catalog
EMAIL:
ken.davidson@circellar.com
HTTP://www.micromint.com/
RATING: (8)

Micron Semiconductor, Inc. & Subsidiaries
Very wide-ranging manufacturer
including RAM, computers,
displays.
INFO: Extensive product support
and datasheets
EMAIL: webmaster@micron.com
HTTP://www.micron.com/
RATING: (10)

Micronetics (Wireless)
EMAIL: micrnet@aol.com

Micropolis (S) Pte Ltd.
Hard drive manufacturer
INFO: Product specs and config
info
EMAIL:
webmaster@micropolis.com
HTTP://www.micropolis.com/
RATING: (7)

Microsemi Corp.
Semiconductor manufacturer
INFO: Products, support, news,
FAQ. Great EE site!
EMAIL: See site for form
HTTP://www.microsemi.com/
RATING: (10)

Microsoft Corporation
Software, O/S company

INFO: Products, search, support, shop, you name it!
EMAIL: See site for forms
HTTP://www.microsoft.com/
RATING: (10)

Microstar Laboratories, Inc.
Data acquisition and control products
INFO: Apps, downloadable software, documents
EMAIL: info@mstarlabs.com
HTTP://www.mstarlabs.com/
RATING: (9)

Microtec
Embedded software tools & real-time operating systems
INFO: News, jobs, articles, FAQs, tech, training
EMAIL: info@mri.com
HTTP://www.mri.com/
RATING: (9)

Microtest, Inc.
CD ROM, CD recordables, scanners, network test
INFO: Demos, news, tech, drivers
EMAIL: info@microtest.com
HTTP://www.microtest.com/
RATING: (8)

Microwave Communications Laboratories, Inc.
Manufacture RF & microwave components
INFO: Product info/specs
EMAIL: mcli@mcli.com
HTTP://www.mcli.com/
RATING: (8)

Microwave Distributors
Distributes microwave & RF components
INFO: Linecard, inventory, news
EMAIL: See site for form
HTTP://www.microwavedistributors.com/
RATING: (9)

Microwave Journal Magazine
Online version of magazine
INFO: This month's issue, new products, subscribe, online manufacturer's product directory
EMAIL: mwj@mwjournal.com
HTTP://www.mwjournal.com/mwj.html
RATING: (9)

Midcom, Inc.
Manufactures transformers
INFO: Online help, apps, design articles database, links — Great source for tech articles.
EMAIL: sales@midcom.anza.com
HTTP://www.midcom-inc.com/
RATING: (10)

Milgray Electronics, Inc.
Industrial components distributor
INFO: Linecard, locations, specials, links
EMAIL: info@milgray.e-mail.com
HTTP://www.milgray.com/
RATING: (8)

Mill-Max Mfg.
Interconnectors & sockets
INFO: Products, tech, distributors
EMAIL: millmax@garvan.com
HTTP://www.mill-max.com/
RATING: (8)

Millennium Antenna
PCB antennas design/manufacturing

M

INFO: Product/service info
EMAIL: milantco@borg.com
HTTP://www.borg.com/~milantco/
RATING: (7)

Miller Freeman plc

Business-business publishers
INFO: Electronics related news in
UK, publishers info
EMAIL: ed98@cityscape.co.uk
HTTP://www.dotelectronics.co.uk/
RATING: (9)

Miniature Precision Components

Linear motion components
distributor
INFO: Linecard, catalog
EMAIL:
sales@mpcomponents.com
HTTP://www.mpcomponents.com/
RATING: (7)

MIPS Technologies, Inc.

RISC MCU, MPU manufacturer
INFO: Suppliers, development
tools, news, jobs
EMAIL: See site for form
HTTP://www.mips.com/
RATING: (9)

Mirage Electronics

Distributes semiconductors
worldwide
INFO: Stock list, inventory, links,
purchase stock
EMAIL: See site for form
HTTP://www.mirageus.com/
RATING: (10)

Mission Electronics

Semiconductor brokerage firm
INFO: Company/product info
EMAIL: See site for listing
HTTP://www.dram.com/
RATING: (7)

Mitel Corp — Semiconductor Div.

Semiconductor manufacturer
INFO: Sales, support, contact, apps,
datasheets
EMAIL: See site for locations
listing
HTTP://www.semicon.mitel.com/
RATING: (9)

Mitron Corp.

Manufacturing automation
software/services
INFO: Service/product info, links,
you name it!
EMAIL: info@mitron.com
HTTP://www.mitron.com/
RATING: (9)

Mitronics, Inc.

Distributor of electronic
components
INFO: Linecard, specials, inventory
EMAIL: mitron@mitronics.com
HTTP://www.mitronics.com
RATING: (8)

Mitsi - Micro Technology Services, Inc.

Engineering/software/
manufacturing
INFO: Product info, bulletin board
EMAIL: sales@mail.mitsi.com
HTTP://www.mitsi.com/
RATING: (7)

Mitsubishi Electronic Corp.
Electronics manufacturer
INFO: R&D, company/product
info, you name it!
EMAIL:
webmaster@melit.melco.co.jp
HTTP://www.melco.co.jp/
HTTP://www.mela-itg.com/
RATING: (10)

Mitsumi Electronics Corp.
Peripheral manufacturer
INFO: Company/product info,
drivers, etc.
EMAIL: n/a
HTTP://www.mitsumi.com/
RATING: (7)

MMC Electronics
SMD products manufacturer
INFO: Product info/specs
EMAIL: sales@mmea.com
HTTP://www.mmea.com/
RATING: (7)

Mobile Satellite Telecommunications Page
Brian McIntosh's page
INFO: Links to mobile satellite
resources
EMAIL:
100120.3702@compuserve.com
HTTP://www.wp.com/
mcintosh_page_o_stuff/tcomm.html
RATING: (8)

Mode Electronics Ltd.
Canadian electronics components
supplier
INFO: Company/product info,
news, distributors
EMAIL: See site for form
HTTP://www.mode-elec.com/
RATING: (7)

Model Technology
HDL simulation software
INFO: Product info, registered
download, etc.
EMAIL: greg_seltzer@model.com
HTTP://www.model.com/
RATING: (7)

Molex, Inc.
Manufactures connectors/sockets/
cables
INFO: Product info/select, jobs,
news
EMAIL: amerinfo@molex.com
HTTP://www.molex.com/
RATING: (8)

Mondotronics' Robot Store
Robotic supplies for sale
INFO: Catalog, resources, links
EMAIL: info@mondo.com
HTTP://www.robotstore.com/
RATING: (10)

Morgan Kaufmann Publishers, Inc.
Advanced computer science &
engineering books
INFO: Book descriptions, tables of
contents, downloadable
supplements, bookstore listing
EMAIL: mkp@mkp.com
HTTP://www.mkp.com
RATING: (8)

Mosel-Vitelic
RAM manufacturer
INFO: News, support, sales
channels, datasheets
EMAIL: webmaster@mosel-
vitelic.com

M

HTTP://www.moselvitelic.com/
RATING: (9)

Mosis
VLSI fabrication service
INFO: Everything imaginable to
help designing VLSI
EMAIL: mosis-help@mosis.org
HTTP://www.isi.edu/mosis/
RATING: (10)

Motorola, Inc.
Semiconductor and other products
manufacturer
INFO: Information supersite!
EMAIL: See individual divisions
for addresses
HTTP://www.mot.com/
RATING: (Over-all: 10)*****

Motorola High Performance Embedded Systems Division
HTTP://www.mot.com/hpesd/

Motorola Information Systems Group
HTTP://www.mot.com/mims/isg/

Motorola Microcontroller Chips
HTTP://freeware.aus.sps.mot.com/

Motorola Online Data Library
HTTP://www.motserv.indirect.com/

Motorola Semiconductor Products Design-NET
HTTP://motserv.indirect.com/
HTTP://www.design-net.com/

Motorola World-Wide-Web Server
HTTP://www.motorola.com/

Mountain Technologies Ltd.
Cable assembly manufacturer
INFO: Product info, sales, news
EMAIL: abuttjes@mtntech.com
(Pres.)
HTTP://www.mtntech.com/
RATING: (7)

Mouser Electronics, Inc.
Mail order electronics parts
INFO: Sales, tech, linecard
EMAIL: catalog@mouser.com
HTTP://www.mouser.com
RATING: (7)

MPM Corp.
Precision deposition systems for
assembly
INFO: Company, product,
locations, training
EMAIL: lkane@mpmcorp.com
HTTP://www.mpmcorp.com/
RATING: (8)

Murphy Electronics, Inc.
Distributor of passives and
interconnects
INFO: Linecard, hints, links
EMAIL:
murphyelec@murphyelec.com
HTTP://www.murphyelec.com/
RATING: (7)

Mushroom Components
UK importer distributor
INFO: Stock, wanted, news, quiz
EMAIL: pv@mushroom.co.uk
HTTP://www.mushroom.co.uk/
RATING: (7)

Music Semiconductor, Inc.
Memory and related chips
INFO: Products, CAM info,
literature, links
EMAIL: info@music.com
HTTP://www.music.com/
RATING: (9)

MWK Industries
Laser supplies
INFO: Product info, FAQs
EMAIL: mkenny1989@aol.com
HTTP://www.pweb.com/mwk/
main.htm
RATING: (8)

Nanothinc
Nanotechnology publisher/
developer/distributor/etc.
INFO: NanoNews, NanoMarkets,
NanoWorld, NanoSCi. The
EVERYTHING site for
nanotechnology!
EMAIL:
webmaster@nanothinc.com
HTTP://www.nanothinc.com/
RATING: (10)*****

National Electronics Service Dealers Association, Inc. (NESDA)
Serves the professional interests of
the independent segments of the
electronics service industry.
INFO: Divisions, activities,
legislation, member discounts,
certification, newsletter, magazine,
etc.
EMAIL: clydenesda@aol.com
HTTP://www.nesda.com
RATING: (n/a)

National Institute of Standards and Technology
NIST
INFO: General info, search,
programs, links, FAQs
EMAIL: See site for staff directory
HTTP://www.nist.gov:80/
RATING: (10)

National Instruments
Instrumentation hardware/software
INFO: Catalog, support,
newsletters, great data for
instrumentation in general
EMAIL: info@natinst.com
HTTP://www.natinst.com
RATING: (10)

National Semiconductor Corp.
Semiconductor manufacturer
INFO: Extensive company and
product info with datasheets,
specs, apps
EMAIL: Onsite feedback form
HTTP://www.nsc.com/
RATING: (10) *****

NEC Electronics, Inc.
Semiconductors, consumer
electronics, etc.
INFO: Products, links, huge
database, support
EMAIL: tech-support@nec.com
HTTP://www.nec.com/
RATING: (10)

Needham's Electronics, Inc.
Device programmers manufacturer/
retailer
INFO: Device info, latest software,
pricing, FAQs

N

EMAIL: See site
HTTP://www.needhams.com/
FTP://ftp.needhams.com/pub/users/
needhams
RATING: (8)

Nemal Electronics International, Inc.
Cable and connector manufacturer, distributor
INFO: Product info, news
EMAIL: nemal@internetco.net
HTTP://www.nemal.com/
RATING: (7)

Nepenthe Distribution, Inc.
Manufacturer/distributor of interconnectors
INFO: Product info, linecard, datasheets
EMAIL: sales@nepen.com
HTTP://www.nepen.com/
RATING: (8)

NetGuide(tm) Magazine
Online magazine for Net news and hardware
INFO: Useful to electronics as Net hardware info
EMAIL: See site for form
HTTP://techweb.cmp.com/net/
current/
RATING: (10)

The Network
Electronics buyer directory
INFO: Links to distributors, reps, manufacturers
EMAIL: webmaster@electronet.com
HTTP://www.electronet.com/
RATING: (7)

Neumann Electronics, Inc.
Distributing passives & interconnectors
INFO: Linecard, company info
EMAIL: info@neumann-elec.com
HTTP://www.neumann-elec.com/
RATING: (7)

New England Circuit Sales (NECX)
Semiconductor/computer components distributor
INFO: Online shopping
EMAIL: See site for form
HTTP://www.necx.com/
RATING: (6)

Newbridge Networks Corp.
Network products manufacturer
INFO: Product info, user groups, news, help files, you name it!
EMAIL: webmaster@newbridge.com
HTTP://www.newbridge.com/
RATING: (10)

Newport Components Ltd.
UK power converter manufacturer
INFO: DC-DC data, info, sales
EMAIL: techsupp@newport.co.uk
HTTP://www.newport.co.uk/
RATING: (7)

Newton's International Electrical Journal
Online journal by Gerald C. Newton.
INFO: Ask electrical questions, articles, book list, slide shows
EMAIL: See site for form
HTTP://www.polarnet.com/users/
gnewton/newt.htm
RATING: (10)

NexGen
SEE: *Advanced Micro Devices, Inc.*

NeXT Software, Inc.
Computer company
INFO: WebObjects, services,
product info
EMAIL: international@next.com
HTTP://www.next.com/
RATING: (9)

NIC Components Corp.
Manufactures passive components
INFO: Product info/apps,
distributors/reps, links
EMAIL: sales@niccomp.com
HTTP://www.niccomp.com/
RATING: (9)

Nikola Tesla Information Page
Bill Beaty's homepage
INFO: Links, info on Tesla science/
coils
EMAIL: billb@eskimo.com
HTTP://www.eskimo.com/~billb/
tesla/tesla.html
RATING: (8)

Nohau Corp.
In-circuit emulators
INFO: Microcontrollers supported,
with specs
EMAIL: support@nohau.com
HTTP://www.nohau.com
RATING: (9)

Norham Radio, Inc.
Canadian amateur radio dealer
INFO: Specials, links, inventory
EMAIL: norham@fox.nstn.ca
HTTP://www.norham.com/
RATING: (8)

North American Capacitor Company (NACC)
Mallory products
INFO: Inventory, contests, sounds,
help, you name it!
EMAIL: See site for form
HTTP://www.nacc-mallory.com/
RATING: (9)

North Atlantic Components, Inc.
Independent stocking distributor
INFO: Linecard, links
EMAIL: buyers@northatlantic.com
HTTP://www.northatlantic.com/
RATING: (7)

Norvell Electronics, Inc.
Electronic components distributor
INFO: Offices, linecard, info
request
EMAIL: business@norvell.com
HTTP://www.norvell.com/
RATING: (8)

Novalog, Inc.
IR products (IrDAs) manufacturer
INFO: Datasheets, links, reps
EMAIL: jchimo@novalog.com
HTTP://www.novalog.com/
RATING: (7)

NTE Electronics, Inc.
Manufactures semiconductors/
passives
INFO: Product info/FAQs/specs,
distributors, download cross ref.
EMAIL: maillist@nteinc.com
HTTP://www.nteinc.com/
RATING: (10)

N - O

Nu-Way Electronics Distributor
INFO: Linecard
EMAIL:
102443.641@compuserve.com
HTTP://www.electronet.com/
nuway.htm
RATING: (2)

Nu Horizons Electronics Corp.
Semiconductor distributor
INFO: Linecard, datasheets, tech
papers, etc.
EMAIL: info@nuhorizons.com
HTTP://www.nuhorizons.com/
RATING: (10)

Nutron Computers and Electronics, Inc.
Retail/wholesale components
distributor
INFO: Online catalog and ordering
EMAIL: nutron@gonutron.com
HTTP://www.gonutron.com/
RATING: (8)

Nuts & Volts Magazine
Exploring Electronics and
Technology
INFO: Files, issues, stamp
applications, subsriptions, etc.
EMAIL: editor@nutsvolts.com
HTTP://www.nutsvolts.com/
FTP://ftp.nutsvolts.com/pub/
nutsvolts/library
RATING: (10)

NVE (Nonvolatile Electronics, Inc.)
Nonvolatile manufacturer including
MRAM and GMR sensors

INFO: Product info, datasheets,
applications notes
EMAIL: info@nve.com
HTTP://www.nve.com/
RATING: (8)

NVidia Corp.
Multimedia accelerator chips
INFO: FAQs, software upgrades,
data
EMAIL: info@nvidia.com
HTTP://www.nvidia.com/
RATING: (10)

Oak Frequency Control
Manufacturer of frequency control
products
INFO: Catalog, features, reps
EMAIL: techlink@ofc.com
HTTP://www.ofc.com/
RATING: (7)

OakGrigsby
Manufactures switches, solenoids,
etc.
HTTP://www.industry.net/
oakgrigsby

Oatley Electronics
Kits, supplies from Australia
INFO: Catalog, contact
EMAIL: Oatley@world.net
HTTP://www.ozemail.com.au/
~oatley/
RATING: (6)

ODU-USA
Connects and contacts
manufacturer
INFO: New, products/company
info/support
EMAIL: ralfeb@odu-usa.com
HTTP://www.odu-usa.com/
RATING: (7)

O

Ohmite
Developer of passive components
INFO: Product guide, reps, new
EMAIL: ohmite@wwa.com
HTTP://www.ohmite.com/
RATING: (8)

Okaya Electric America, Inc.
Noise supressors, PC104s, displays
INFO: Product info/specs, reps
EMAIL: n/a
HTTP://www.okaya.com/
RATING: (7)

OKI Semiconductor, Inc.
Semiconductor manufacturer
INFO: Products info/database, jobs,
contact
EMAIL: See site for reps
HTTP://www.okisemi.com/
HTTP://www.oki.co.jp/
RATING: (9)

Olflex Wire & Cable
EMAIL: sales@olflex.com
HTTP://www.olflex.com/
RATING: (3)

Omega Engineering, Inc.
Process measurement and control
INFO: New, software, tech
references, wonderful electronics
references on this site
EMAIL: See site for form
HTTP://www.omega.com/
RATING: (10) ***

Omron Electronics, Inc.
Factory automation, control
components
INFO: Product, support,
distributors, tech
EMAIL:
webmaster@oei.omron.com

HTTP://www.oei.omron.com/
RATING: (8)

OnLine Electronic Publishing
Online publication collection
INFO: Online books
EMAIL: jlutgen@earthlink.net
HTTP://home.earthlink.net/
~jlutgen/epublish.html
RATING: (9)

Online Technology Exchange
Obsolete, hard-to-find, and military
components
INFO: Inventory search, sales, links
EMAIL:
webmaster@onlinetechx.com
HTTP://www.onlinetechx.com/
RATING: (9)

ONSALE - Auctions of Computers and Electronics
Live online auction house
INFO: Current bids and winners
EMAIL: See site
HTTP://www.onsale.com
RATING: (8)

OnSpec Oscillators Limited — UK
Manufactures custom oscillators
INFO: Specs, design, production
EMAIL: sales@onspec.co.uk
HTTP://www.onspec.co.uk/
RATING: (9)

Opti, Inc.
Multimedia chip sets and other
silicon solutions
INFO: Products, jobs, overview,
FAQs

O

EMAIL: webmaster@opti.com
HTTP://www.opti.com/
FTP://ftp.opti.com/
RATING: (7)

Opto22
Manufactures solid-state relays
HTTP://www.industry.net/opto22

Optotek, Ltd.
High-frequency CAE/CAT
software, LED products
INFO: Product info, reps. Cute site!
EMAIL: optotek@optotek.com
HTTP://www.optotek.com
RATING: (9)

Orbit Semiconductor, Inc.
Manufactures gate arrays
INFO: Products, sales, jobs
EMAIL: sales@orbitsemi.com
HTTP://www.orbitsemi.com/
RATING: (8)

OrCAD, Inc.
EDA software vendor
INFO: Demos, FAQs, reps, libraries
EMAIL: info@orcad.com
HTTP://www.orcad.com
FTP://ftp.orcad.com/
RATING: (10)

Oriental Motor
Small motor manufacturer
HTTP://www.industry.net/
oriental.motor

Orion Instruments, Inc.
In-circuit emulators, uC software/
developers, etc.
INFO: Products, tech support,
news, jobs, links, FAQs

EMAIL: info@oritools.com
HTTP://www.oritools.com/
FTP://ftp.oritools.com
LISTSERV: lists@oritools.com
with 'join announce' in msg body
RATING: (10)

Ormix — with Microchip PIC Info Page
Import/export in Russia, etc.
INFO: General, Microchip info,
offices
EMAIL: ormix@mail.ormix.riga.lv
HTTP://www.ormix.riga.lv/
RATING: (8)

Oscillatek
Quartz crystal oscillators
INFO: Product, contact, links
EMAIL: sales@otek.com
HTTP://www.otek.com/
RATING: (9)

OSU Department of Computer and Information Science
HTTP://www.cis.ohio-state.edu:80/
hypertext/faq/usenet/FAQ-List.html
SEE: *List of USENET FAQs*

Oxford Semiconductor, Ltd.
Designs, manufactures ASICs
INFO: Design service, design flow,
etc.
EMAIL: See site for form
HTTP://www.alt-tech.com/
products/oxford/
RATING: (6)

Oxford Computer, Inc.
EMAIL:
oxfordcomput@delphi.com

Oz Technologies, Inc.
EMAIL: oztech@ix.netcom.com

P & D Electronics
Electronics parts brokers
INFO: Linecard
EMAIL: n/a
HTTP://www.pnde.com/
RATING: (2)

Pace Electronics, Inc.
Group of 3 companies
INFO: Inventory,specials, contact
EMAIL: info@pace-net.com
HTTP://www.pace-net.com/
RATING: (7)

Pacific Radio Electronics
Pro audio/video broadcast supply
distributor
INFO: Specials, catalog, links
EMAIL: See site for listing
HTTP://www.pacrad.com/
RATING: (7)

Pacific Research Technology, Inc.
OEM buying service for
components
INFO: Locations, requests, sales,
stock, specials
EMAIL: sales@pacificresearch.com
HTTP://www.pacificresearch.com/
RATING: (8)

Pacific Semiconductor, Inc.
Import/export wholesale of
semiconductors
INFO: Contact, specials, inventory
in Canadian & US $s, links
EMAIL: info@pacificsemi.com
HTTP://www.pacificsemi.com/
RATING: (10)

Packard Hughes Interconnect
A GM subsidiary handling
connectors, fiber optics, etc.
INFO: Products, jobs, request form
EMAIL: See site for form
HTTP://www.packardhughes.com/
RATING: (7)

PADS Software, Inc.
EDA software
INFO: FAQs, product info, news,
EDA info
EMAIL: web@pads.com
HTTP://www.pads.com/
RATING: (10)

Page-O-Links
EE sites
INFO: Many links to EE sites
EMAIL: coufal@vvm.com
HTTP://www.vvm.com/~coufal/
RATING: (8)

PAiA Electronics, Inc.
Hobbyist kits
INFO: Catalog, contests, FAQs,
links
EMAIL: paia@paia.com
HTTP://www.paia.com/
RATING: (9)

Panamax
Power protection products
INFO: Product/company info,
specs, advice, fun
EMAIL: marcom@panamax.com
HTTP://www.hooked.net/panamax/
RATING: (9)

Panduit Corporation
Wiring and communication
products
INFO: Company/product info/apps/
tech

P

EMAIL: info@panduit.com
HTTP//www.panduit.com/
RATING: (8)

Paradigm Technology, Inc.
Develops and produces SRAM mods
INFO: Sales, products, news, datasheets
EMAIL: See site for listing
HTTP://www.prdgm.com/
RATING: (7)

Parallax, Inc.
Basic stamp and PIC programmers
INFO: News, products, support, links, software, apps, notes, you name it!
EMAIL: info@parallaxinc.com
HTTP://parallaxinc.com/
HTTP://www.parallaxinc.com/
FTP://ftp.parallaxinc.com/
RATING: (10)

Parallel and Distributed Systems Group
U of Waterloo E&CE department
INFO: Repository for FPGA, transputer, and other research
EMAIL: wdbishop@dictator.uwaterloo.ca
HTTP://www.pads.uwaterloo.ca/~pads
RATING: (9)

PartNet
The parts information network
INFO: Mech/Elec. engineers, search over 1,000,000 electrical, electro-mechanical and mechanical parts from various suppliers worldwide. Also, view data sheets, CAD models, pricing, availability, etc. GREAT TOOL!
EMAIL: webmaster@part.net
HTTP://part.net/partnet/
RATING: (10) **********

Paul Maxwell-King Homepage and Related Pages
INFO: Valuable projects, hacks, links for PICs
EMAIL: paul@maxking.demon.co.uk
HTTP://www.gpl.net/paulmax/
RATING: (10)

PC Engines
Consulting in uCs, uPs, etc.
INFO: Simms/electronic component help files. Excellent resources for designers of uC/uP products or anyone for that matter.
EMAIL: pdornier@best.com
HTTP://www.best.com/~pdornier/
RATING: (10)

PCA Electronics, Inc.
Manufacturer of wire-wound electro-magnetic components
INFO: Contact, product info/specs
EMAIL: See site for list
HTTP://www.pcainc.com/
RATING: (7)

PCIM Online
Power conversion & intelligent motion
INFO: Articles, buyers guide, conferences
EMAIL: editor@pcim.com
HTTP://www.pcim.com/
RATING: (10)

Peerless Electronic Supplies
Distributor of components/ equipment

INFO: Specials, links, linecard
EMAIL: sales@pelec.com
HTTP://www.pelec.com/
RATING: (7)

Peerless Radio Corp.
Electromechanical/interconnect
products
INFO: Linecard, links, contact
EMAIL: Only request form on site
HTTP://www.peerless-rc.com/
RATING: (7)

PennWell Publishing Co.
Advanced technology division
See site for listing of online
magazines
INFO: Links to individual sites
EMAIL: jeffd@pennwell.com
HTTP://www.lfw.com/
RATING: overall (10)

Penstock — An Avnet Company
Distibutor of RF and microwave
semiconductors
INFO: Linecard, history, product,
contact
EMAIL:
penstock.comments@avnet.com
HTTP://www.penstock.avnet.com/
RATING: (7)

Pericom Semiconductor Corp.
INFO: Order book, product info/
specs.
EMAIL: nolimits@pericom.com
HTTP://www.pericom.com/
RATING: (9)

Periphex, Inc.
Replacement batteries manufacturer
INFO: Dealers, products with costs
EMAIL: periphex@aol.com
HTTP://home.navisoft.com/
periphex/
RATING: (6)

Personal Engineering & Instrumentation News
INFO: Articles for electrical/
electronic engineers
EMAIL: See site for form
HTTP://www.pein.com/
RATING: (8)

Phihong USA
Power solutions manufacturer
INFO: Distributors, reps, contacts
EMAIL: usasales@phihongusa.com
HTTP://www.phihongusa.com/
RATING: (5)

Philips Semiconductors
Semiconductor manufacturer
INFO: Products, news, tech data,
jobs, distributors
EMAIL:
webmaster@www.semiconductors.
philips.com
HTTP://
www.semiconductors.philips.com/
RATING: (10)

Phoenix Contact
Manufactures terminal blocks,
modules
HTTP://www.industry.net/
phoenix.contact

Phoenix Enterprises
Distributor specializing in
connectors/sockets
INFO: Product pics, linecard
EMAIL:sales@phoenixent.com

P

HTTP://www.phoenixent.com/
RATING: (7)

PIC Mail Archive
Send and receive mail list for PIC
subjects
INFO: Archive and subscribe/send
instructions
EMAIL: luehrb@cdr.Stanford.edu
HTTP://cdr.stanford.edu/people/
luehrb/up/pic/
LISTSERV: See www site for
instructions
RATING: (9)

PIC Microcontroller FAQ and Links
SEE: *T.A.K. DesignS*

PICCO
VLSI design consultants
INFO: Company profile, services,
chip images
EMAIL: info@picco.com
HTTP://www.picco.com/
RATING: (7)

Pilkington Microelectronics Ltd.
Research and design company
INFO: Free auto layout pgms,
products, support
EMAIL info@pmel.com
HTTP://www.pmel.com/
RATING: (9)

Pipe Thompson
Distributor
EMAIL: pipethom@idirect.com

Pixel Magic, Inc.
EMAIL: hr@pixelmagic.com

Plano Components
Stocking distributor,
semiconductors
INFO: Company info, links, trends
EMAIL: netsales@plano.com
HTTP://www.plano.com/plano
RATING: (7)

Platform Systems, Inc.
Prototyping tools for hobby/
education
INFO:Hints, products, distributors
EMAIL: platform@cadvision.com
HTTP://www.cadvision.com/
platform/
RATING: (7)

PMC-Sierra, Inc.
Network component solutions
INFO: Product documents, jobs,
etc.
EMAIL: info@pmc-sierra.bc.ca
HTTP://pmc-sierra.bc.ca
RATING: (9)

PneuTouch
Custom robotic label applicator/
printers
INFO: Label applications for PCBs
EMAIL: pneutouch@aol.com
HTTP://www.lunaweb.com/
pneutouch
RATING: (7)

PNY Electronics, Inc.
Simms manufacturer
INFO: Distributors, jobs, news, tech
EMAIL: See site for form/listing
HTTP://www.pny.com/
RATING: (9)

Polaris Industries
CCD and other video products
INFO: Products, ordering, etc.

EMAIL: See site
HTTP://www.polarisusa.com/
RATING: (9)

Poly-Flex Circuits, Inc.
Flexible PCB technology
INFO: Company site, apps
EMAIL: pfcsales@ids.net
HTTP://www.polyflex.com/
RATING: (8)

Portland Area Robotics Society
INFO: Club info, links, robotics info/help files
EMAIL: marvin@agora.rdrop.com
HTTP://www.rdrop.com/users/
marvin/
RATING: (9)

Powell Electronics, Inc.
Value-added service distributor
INFO: Linecard, mil-specs, contact
EMAIL: info@powell.com
HTTP://www.powell.com/
RATING: (8)

Power Convertibles Corp.
Power converters
INFO: Online databook, apps, sales offices
EMAIL: info@pcc1.com
HTTP://www.pcc1.com/
RATING: (7)

Power Quality Assurance Magazine
Publisher's Website
INFO: Articles, buyers guide, forum, you name it!
EMAIL: sales@powerquality.com
HTTP://powerquality.com
RATING: (10)

Power Trends, Inc.
Power converters manufacturer
INFO: Product pics/specs
EMAIL: marketing@powertrends.com
HTTP://www.powertrends.com/isr/
RATING: (7)

PowerKit
Distributor of power supplies and related
INFO: Order catalog, etc.
EMAIL: powerkt@gramercy.ios.com
HTTP://gramercy.ios.com/
~powerkt/
RATING: (5)

Powersonic Corp.
Sealed rechargeable batteries
INFO: Apps, sales, product/company info
EMAIL: battery@power-sonic.com
HTTP://www.power-sonic.com/
RATING: (7)

Precision Resistive Products, Inc.
Manufactures metal film resistors
INFO: Product info/specs, reps
EMAIL: info@prpinc.com
HTTP://www.prpinc.com/
RATING: (7)

Prem Magnetics, Inc.
Transformer and inductor manufacturer
INFO: Product info/specs, new
EMAIL: sales@premmag.com
HTTP://www.premmag.com/
RATING: (8)

P

Premier Farnell
SEE: *Farnell*

Prentice Hall
Publisher
INFO: Books include professional tech refs, engineering, etc. Look up that book you have been wanting!
EMAIL: See site for Email listing and listserv
HTTP://www.prenhall.com/
RATING: (10)

Pride Electronics, Inc.
Industrial imaging distributor
INFO: Linecard with pics
EMAIL: 104320673@compuserve.com
HTTP://www.ibiinc.com/pride/
RATING: (6)

Printed Circuit Design's CyberCafe
A huge site for PCB info/software from PCB Magazine
INFO: News, magazine, resourses, chat, software, classifieds, links
EMAIL: See site
HTTP://www.pcdmag.com/
RATING: (10)

Pro Components, Inc.
Parts broker with four international locations
INFO: Linecard, stock, specials
EMAIL: thepros@earthlink.net
HTTP://www.procomponents.com/
RATING: (8)

Programmable Logic Performance Corp.
PLD design engineering help
INFO: Complete tests for PLDs
EMAIL: info@prep.org
HTTP://www.prep.org/
RATING: (10)

Progressive Image
Components distributor
INFO: Linecard, commodity list, tour
EMAIL: progressive@progressiveimage.com
HTTP://www.progressiveimage.com/
RATING: (7)

Projects Unlimited, Inc.
Manufactures audio transducers, etc
INFO: Product info/specs, articles
EMAIL: sales@pui.com
HTTP://www.pui.com/
RATING: (7)

ProTech Books
Bookstore with computer/technical/electronics books
INFO: Searchable database, buy books, links
EMAIL: store2@pro-tech.com
HTTP://www.pro-tech.com/
RATING: (8)

Protek
SEE: *H C Protek*

Protel Technology, Inc.
Windows-based EDA software
INFO: Product info/apps, help files, demos, links
EMAIL: salesusa@protel.com
HTTP://www.protel.com/
RATING: (10)

Pulse
Magnetic-based products manufacturer

INFO: New, product/company info
EMAIL: custserv@pulseeng.com
HTTP://www.pulseeng.com
RATING: (7)

Q Components
IC/Component distributor
INFO: Full online catalog, hot picks
EMAIL: info@qcomponents.com
HTTP://www.qcomponents.com/
RATING: (8)

Q-Logic Corp.
ICs, adaptor cards (SCSI), etc.
INFO: Sales, products, tech support
EMAIL: g_munck@qlc.com
HTTP://www.qlc.com/
RATING: (8)

Qtec Semiconductor Pty., Ltd.
Provider of IC design services
INFO: Services description
EMAIL: qtec@webventures.com.au
HTTP://www.webventures.com.au/
ElectTech/qtec/
RATING: (5)

Quadrillion Corp.
Yield analysis software for semiconductor manufacturers
INFO: Training info, semiconductor links, etc.
EMAIL: info@quadrillion.com
HTTP://www.quadrillion.com/
RATING: (6)

Qualcomm, Inc., Including Eudora Email Software
Digital wireless technology manufacturer/licenser
INFO: Locations, product info, links, Eudora
EMAIL: See site for info
HTTP://www.qualcomm.com/

HTTP://www.eudora.com/
FTP://ftp.qualcomm.com/
RATING: (10)

Qualitek International, Inc.
Solder supplies manufacturer
INFO: Product info/news, contact
EMAIL: qualitek@ix.netcom.com
HTTP://www.qualitek.com/
RATING: (7)

Quality Semiconductor, Inc. (QSI)
Semiconductor manufacturer, OEM
INFO: Apps, product info, sales, etc.
EMAIL: webmaster@qualitysemi.com
HTTP://www.qualitysemi.com/
RATING: (9)

Quality Technologies Corp.
MCI Mail ID: 4286653

QuestLink Technology, Inc.
Resource for info on ICs
INFO: Search for info/datasheets on tons of ICs, MCUs, MPUs. Wonderful resource site!
EMAIL: See site for form
HTTP://www.questlink.com/
RATING: (10)**********

QuickLogic Corp.
FPGA chips/software manufacturer
INFO: Software, product info/tech
EMAIL: info@qlogic.com
HTTP://www.quicklogic.com/
RATING: (10)

Q - R

Quickturn Design Systems, Inc.
Design verification through logic emulation
INFO: Products/services
EMAIL: webmaster@quickturn.com
HTTP://www.quickturn.com/
RATING: (7)

R & D Electronic Supply
Components distributor
INFO: Linecard
EMAIL: rd_electronics@interfin.com
HTTP://www.interfin.com/r%26delectronics/
RATING: (4)

R. L. Drake
Communications manufacturer
INFO: Satellite/cable products info, tech
EMAIL: See site for list
HTTP://www.rldrake.com/
RATING: (7)

R O Associates, Inc.
Power converter modules manufacturer
INFO: Products, rep links
EMAIL: sales@roassoc.com
HTTP://www.roassoc.com/
RATING: (7)

Racal Instruments, Inc.
Automated test equipment
INFO: Products, press, reps, links
EMAIL: 72056.256@compuserve.com
HTTP://www.racalinst.com/
RATING: (8)

Radio Devices
Dealer of amateur radio products
INFO: Product info, links, radio sites
EMAIL: rhynek@ultranet.com
HTTP://www.ultranet.com/biz/raddev/
RATING: (8)

Radio Electric Supply
Receiving/special-purpose tubes
INFO: Price list, order info
EMAIL: phmd51a@prodigy.com
HTTP://www.mindspring.com/~gies/res.html
RATING: (8)

Radio Shack
SEE: *Tandy Corp.*

RadiSys
Embedded computer manufacturer
INFO: Product info/support, contact, news
EMAIL: See site for directory
HTTP://www.radisys.com/
RATING: (9)

RAF Electronic Hardware
Electronics hardware manufacturer
INFO: Sales, products, literature request
EMAIL: info@rafhdwe.com
HTTP://www.rafhdwe.com/
HTTP://www.fastenermall.com
RATING: (8)

Raltron Electronics Corp.
Crystals, filter, oscillators, etc.
INFO: Product info/specs, profile
EMAIL: sales@raltron.com
HTTP://www.raltron.com/
RATING: (7)

Ramtron International Corp.

Nonvolatile (FRAM) memory
manufacturer
INFO: Products, sales, news, tech,
jobs
EMAIL: info@ramtron.com
HTTP://www.csn.net/ramtron/
RATING: (8)

Raymond Sarrio Company

Ham radio Web store
INFO: Ham radio products, links,
polls. Good Ham stop over!
EMAIL: wb6siv@cyberg8t.com
HTTP://www.sarrio.com/
RATING: (10)

Raytheon Company

Manufactures a range of electronics
products
INFO: See divisions for info
EMAIL: See site for form
HTTP://www.raytheon.com/
RATING: (8)

RCD Components, Inc.

Manufacturer of resistors, coils, etc.
INFO: Online ordering, product/
stock info
EMAIL: rcdcompinc@aol.com
HTTP://www.rcd-comp.com/
RATING: (7)

RdF Corp.

EMAIL: sensor@rdfcorp.com

Real Time Devices, Inc. (RTD)

BBS: 814-234-9427

Real-Time Systems

Spring project — U of
Massachusetts, Amherst
INFO: Real-time data/refs

EMAIL: See site for U listing
HTTP://www-ccs.cs.umass.edu/
RATING: (10)

Reclamation Technologies, Inc.

Asset reclamation for electronics
industry
INFO: Company/service info,
contact
EMAIL: reuse@rectech.com
HTTP://www.rectech.com/
RATING: (8)

Remee Products Corp.

Manufacturer of electronic wire,
fiberoptic cable and assemblies
INFO: Cable industries 1st
downloadable WINDOWS driven
copper/fiber digital catalog
EMAIL: rpcsale@aol.com
HTTP://www.techexpo.com/firms/
remeeprd.html
RATING: (8)

Reptron Electronics

Distribution — 22 locations
INFO: Linecard, inventory search,
vendor links
EMAIL: marcom@reptron.com
HTTP://www.reptron.com
RATING: (7)

Republic Electronics Corp.

Manufactures ceramic capacitors
INFO: Products, reps
EMAIL:
larrye@rec.microserve.com
HTTP://www.repelect.com/
RATING: (7)

R

R

Resistor Color Codes
HTTP://tiger.uofs.edu/department/
psych/resistor.html

Resource 800
Distributor of computer clock
batteries
INFO: Help files/articles, product
info
EMAIL: See site for order-form
HTTP://www.iitexas.com/gpages/
resource.htm
RATING: (8)

The Rework and Repair Exchange
Electronics forum and resource
INFO: Idea exchange, industry
links, software, training and
support; everything you need to
know about soldering and more.
EMAIL: experts@metcal.com
HTTP://www.metcal.com
RATING: (10)

RF Imaging & Communication
Buy/sell pre-owned test equipment
INFO: Items for sale
EMAIL: rfimage@best.com
HTTP://www.best.com/~rfimage/
RATING: (5)

RF Micro Devices, Inc.
Manufacturer of RF ICs
INFO: Datasheets, product info/
specs, reps — great resource for RF
IC info
EMAIL: info@rfmd.com
HTTP://www.rfmd.com/
RATING: (10)

Rhode & Schwarz
Comm. & measuring technology.
INFO: Products, contact, events
EMAIL: See site for listing
HTTP://www.rsd.de/
RATING: (9)

Richard Steven Walz's Almost Gigantic Electronics FTP Resource Site!!
INFO: 800 files relating to
electronics
EMAIL: rstevew@armory.com
HTTP://www.armory.com/~rstevew/
FTP://ftp.armory.com:/pub/user/
rstevew
RATING: (10)

Richardson Electronics, Ltd.
Worldwide electronics distributor
INFO: RF, tubes, power semis
product info/catalog, etc.
EMAIL: Form on site
HTTP://www.rell.com/
RATING: (9)

Richey Electronics
Value-added distributor
INFO: Linecard, services info,
locations
EMAIL: richeydist@richeyelec.com
HTTP://www.richeyelec.com/
RATING: (8)

Robert's Satellite TV Page
Robert Smathers' hobbyist site
INFO: C-band TVRO info, links
EMAIL: roberts@nmia.com
HTTP://www.nmia.com/~roberts/
RATING: (8)

Robotics FAQ — Kevin Dowling
Comp.robotics newsgroup FAQ
INFO: Everything about robotics!

EMAIL: nivek@cmu.edu
HTTP://www.frc.ri.cmu.edu/
robotics-faq/
RATING: (9)

Robotics Institute, Carnegie Mellon University

Compliments of Mike Blackwell
INFO: Index of hundreds of
electronics vendors, listed by
category
EMAIL: mkb@cs.cmu.edu
HTTP://www.frc.ri.cmu.edu/~mkb/
RATING: (9)

Robotics Internet Resources Compendium

Complete robotics Web reference
INFO: Links, add links
EMAIL: dbell@coral.bucknell.edu
HTTP://spectrum.eg.bucknell.edu/
~robotics/rirc.html
RATING: (10)

Robotics Internet Resources Page

UMass Laboratory for Perceptual
Robotics
INFO: Complete robotics internet
resources including FTPs,
demos, circuits, links, FAQs, you
name it!
EMAIL: connolly@ai.sri.com
HTTP://piglet.cs.umass.edu:4321/
robotics.html
RATING: (10)

Robotics Parts from Robert L. Doerr

INFO: Heathkit robot info, robot
parts, etc.
EMAIL: rdoerr@bizserve.com
HTTP://bizServe.com/home/
rdoerr.html
RATING: (7)

R

Rockwell International — Semiconductor Systems

Semiconductor manufacturer
INFO: News, contact, tech, FAQs,
you name it!
HTTP://www.rockwell.com/
HTTP://www.nb.rockwell.com/
RATING: (10)

Ross Technology, Inc.

Subsidiary of Fujitsu Ltd.
Suppliers of SPARC
microprocessors
INFO: Products, sales, support
EMAIL: info@ross.com
HTTP://www.ross.com/
RATING: (9)

RP Electronics Components Ltd.

Components & test equipment
distributor
INFO: Linecard with pics of
products
EMAIL: info@rpelec.com
HTTP://www.rpelec.com/
RATING: (7)

RTG, Inc.

EMAIL: sales@rtg.com

RTMX, Inc.

Real-time O/Ss & device drivers
INFO: Links, O/S info
EMAIL: info@rtmx.com
HTTP://www.rtmx.com/
FTP://ftp.rtmx.com/pub/
RATING: (8)

Russ Hersch's 8051 FAQ

INFO: FAQ for the 8051 line of uCs
EMAIL: russ@shani.net

R - S

USENET: news.answers,
comp.arch.embedded
RATING FOR FAQ rating (10)

S-MOS Systems, Inc.
(A Seiko-Epson Affiliate)
Produces custom & semicustom ICs
INFO: Custom/standard products
info, etc.
EMAIL: webmaster@smos.com
HTTP://www.smos.com/
RATING: (9)

Sage EDA Corp.
EDA/PCB design software
INFO: Essays, product/sales info,
FAQs, demos
EMAIL: support@sage-eda.com
HTTP://www.sage-eda.com/
RATING: (7)

Sager Electronics
National distributor
INFO: Stock checking/pricing,
links, etc.
EMAIL: sagerinfo@sager.com
HTTP://www.sager.com/
RATING: (7)

SAIA-Burgess Electronics — Switzerland
Manufactures switches, motors,
controllers
INFO: Product/sales info/pics
EMAIL: See site for listing
HTTP://www.saia-burgess.com/
RATING: (8)

Sam Engstrom's Electronics Page
Electronics homepage
INFO: Electronics links, PIC stuff,
CMOS datasheets, and other
helpful features. True information
site!
EMAIL: seng@elixir.org
HTTP://www.elixir.org/users/seng
RATING: (10)

Samsung Electronics Co., Ltd.
INFO: Product info/support/
resources, jobs, links to other
Samsung sites, BBS, and many
more!
EMAIL: See site for company-
comm site
HTTP://www.sec.samsung.co.kr/
HTTP://www.samsung.com
RATING: (10)***

Samtec, Inc.
Manufacturer of PCB
interconnections
INFO: Help, samples, tech, links,
design help
EMAIL: info@samtec.com
HTTP://www.samtec.com/
RATING: (9)

Sanbor Corp.
OEM/Consumer cable assembly
house
INFO: Online catalog, OEM info,
tech support
EMAIL: info@sanbor.com
HTTP://www.sanbor.com/
RATING: (10)

Sanders Media Adventures, Inc.
SEE ALSO: *C-Mac Industries, Inc.*
Designs thick-film active video
filters, modules
INFO: Product info/datasheets
EMAIL: wnj_sma@ix.netcom.com

S

SaRonix
Manufactures frequency control
products
INFO: Product info/specs, form for
info
EMAIL: saronix@connectinc.com
HTTP://www.saronix.com/
RATING: (7)

Satellite One
Satellite television retailer/mail
order
INFO: Descriptions, newsletters,
FAQs, etc.
EMAIL: stevenson@netins.net
HTTP://www.netins.net/showcase/
satone
RATING: (9)

Sav-On Electronics
Home electronics, parts, test equip.,
etc.
INFO: Linecard, specials
EMAIL: savon@edm.net
HTTP://www.savon-
electronics.com/
RATING: (6)

Sayal Electronics
Electronics parts seller/buyer
INFO: Specials, inventory, send list
EMAIL: sales@sayal.com
HTTP://www.sayal.com/
RATING: (8)

SB Electronics, Inc.
Manufactures capacitors
INFO: Linecard, specs, reps, etc.
EMAIL: sbe@internetmci.com
HTTP://www.to-be.com/sbe
RATING: (9)

SBC Computer & Components
Exporting electronics for industry
INFO: Linecard, jokes, company
info, interesting links to
live-camshots all over the world
EMAIL:
sbccomp@sbccomputer.com
HTTP://www.sbccomputer.com/
RATING: (10) ***

SBS Direct
Buyers' group for independent
servicers/retailers
INFO: Service repair personnel data
as well as retail info
EMAIL: michelle@friendly-
net.com
HTTP://www.sbsdirect.com/
RATING: (7)

Schaffner-EMC, Inc.
EMC components, test equipment
INFO: Electro-Magnetic
Compatibity product info, tech,
support, etc.
EMAIL:
102646.3377@compuserve.com
HTTP://www.schaffner.com/
RATING: (8)

Schroff, Inc.
Enclosures & microcomputer
packaging
INFO: Company, products,
liturature
EMAIL: See map for nearest rep.
HTTP://www.schroffus.com/
RATING: (8)

Schurter, Inc.
Manufactures switches, fuses,
indicators, etc.

S

INFO: Product datasheets, reps
EMAIL: See site for rep.
HTTP://www.schurterinc.com/
RATING: (7)

Schuster Electronics
Distributes electromechanical components
INFO: News, linecard, locations
EMAIL: relaydog@aol.com
HTTP://www.schusterusa.com/
RATING: (7)

sci.electronics.* info
SEE: *Filip G.*

Scitex Digital Video
Digital video & editing
INFO: Product info/support/demos
HTTP://www.abekas.com/
RATING: (10)

Scope Systems
Industrial electronic repair & service
INFO: Links, service info, misc.
EMAIL: scope@scopesys.com
HTTP://www.scopesys.com/
RATING: (6)

Scott Edward Electronics
Basic stamp master!
EMAIL: 72037.2612@compuserve.com
HTTP://www.tinaja.com/
HTTP://www.nutsvolts.com/ for article reprints

Scott Electronics Supply
Electronic supply company
INFO: Specials, linecard, service center

EMAIL: info@scottele.com
HTTP://www.scottele.com/
RATING: (8)

SEEQ Technology, Inc.
Data communications semiconductor manufacturer
INFO: Product info/datasheets, sales
EMAIL: See site for listing
HTTP://www.seeq.com/
RATING: (8)

SEI
European Marshall affiliate — SEE *Marshall Industries*
INFO: Links, company datasheets, sales
EMAIL: See feedback form on site
HTTP://www.sei-usa.com/
HTTP://www.sei-europe.com/
RATING: (9)

Seiko Instruments USA, Inc.
ICs, LCDs, fiberoptic components
INFO: News, products/support/ apps, reps
EMAIL: ecd.info@seiko-la.com
HTTP://www.seiko-usa-ecd.com/
RATING: (8)

Selectronics, Inc.
Passives/semiconductors distributor
INFO: Purchasing, deals, inventory
EMAIL: selectro@pond.com
HTTP://selectro.com/
RATING: (7)

Semicoa Semiconductor
Manufactures transistors/ photodiodes, etc.
INFO: Regular/Custom product info/specs
EMAIL: sales@semicoa.com

HTTP://semicoa.com/
RATING: (7)

The Semiconductor Subway
Links to semiconductor sites
INFO: Links to research,
fabrication facilities, etc.
EMAIL: boning@mtl.mit.edu.
HTTP://www-mtl.mit.edu/
semisubway.html
RATING: (10)

Semitech, Inc.
ICs & passives distributor
INFO: Inventory, linecard, links
EMAIL: sales@semitech-inc.com
HTTP://semitech-inc.com/
RATING: (9)

SEMS Electronics
Manufactures electronics devices
INFO: Product info/specs, graphics
and software. Simple, well done
site!
EMAIL: seng@elixir.org
HTTP://www.elixir.org/
SEMS_electronics
RATING: (10)

Semtech Corpus Christie
EMAIL: npsmtchad@aol.com

Servo To Go, Inc.
Manufactures low-cost motion
control boards for PCs
INFO: Specs, links, files to
download, etc.
EMAIL: servotogo@msn.com
HTTP://www.iglou.com/servotogo/
RATING: (9)

SGS-Thomson Microelectronics
Semiconductor manufacturer
INFO: Company news/info,
datasheet library
EMAIL: See site
HTTP://www.st.com/
RATING: (8)

Shallco, Inc.
Manufactures switches, attenuators,
etc.
INFO: Product specs/pics, contact
EMAIL: pdorman@nando.net
HTTP://www.shallco.com/shallco
RATING: (7)

Sharp Electronics Corp.
Manufacturer of consumer
microelectronics
INFO: A near-perfect website with
great info/graphics
direction
EMAIL: See site form
HTTP://www.sharp-usa.com/
RATING: (10)

Shortwave/Radio Catalog
Peter Costello's Radio Related Web
Site
INFO: Radio Links/FAQs/database.
A must for
shortwave/ham/scanner/radio info.
EMAIL: pec@ios.com
HTTP://itre.ncsu.edu/radio/
RATING: (10)

Shure Brothers, Inc.
Microphones, headsets, etc.
INFO: Catalog, contact, pics
EMAIL: sales@shure.com
HTTP://www.shure.com/
RATING: (9)

S

Siemens USA
Electronics technology leader
INFO: News, product database,
contact, links to Siemens' sites
EMAIL: See site for listing
HTTP://www.siemens.com/
RATING: (10)

Sierra Components, Inc.
Semiconductor components
distributor in die form
INFO: Linecard, inventory database
EMAIL: scitahoe@sierracomp.com
HTTP://www.sierracomp.com/
RATING: (7)

Sierra Research & Technology, Inc.
EMAIL: cores@srti.com

Signal Processing Technologies, Inc. (SPT)
Manufactures data conversion &
signal conditioning ICs
INFO: Product info/datasheets,
news, articles. Enjoyable site!
EMAIL: custserv@spt.com
HTTP://www.spt.com/
RATING: (10)

Signal Technology Corp.
Manufactures a range of electronics
components, systems, etc.
INFO: 5 operational locations,
product info available in "pdf"
format
EMAIL: esales@sigtech.com
HTTP://www.sigtech.com/
RATING: (7)

SIGNALWARE Corporation
Design, S/W, H/W consulting for
DSPs
INFO: Product/service info, FAQs
for DSPs, etc.
EMAIL:
signalware@mindspring.com
HTTP://www.hitek.com/signalware/
RATING: (8)

Signetics
SEE: *Philips Semiconductors*

Silicon Graphics, Inc.
Computer manufacturer
INFO: Global sites, tech/
developers, "Serious Fun"
EMAIL: webmaster@sgi.com
HTTP://www.sgi.com/
RATING: (10)

Silicon Studio, Ltd.
uC products/software/services
INFO: Products, files and
schematics
EMAIL: info@sistudio.com
HTTP://www.sistudio.com/
RATING: (10)

Silicon Systems, Inc. (SSI)
(A TI Group)
ASICs manufacturer
INFO: Sales, jobs, product info/data
EMAIL: info@ssil.com
HTTP://www.ssi1.com/
RATING: (10)

Silicon Valley Group, Inc. (SVG)
Supplier of automated wafer
processing equipment
INFO: Company/technology info
EMAIL: n/a
HTTP://www.svg.com
RATING: (7)

SiliconSoft, Inc.
Data acquisition hardware/software for PCs
INFO: Links, files, plenty of great stuff!
EMAIL: foletta@ix.netcom.com
HTTP://www.siliconsoft.com/
RATING: (10)

Simcona Electronics Corp.
Wire/data-comm products distributor
INFO: Products, value added service
EMAIL: info4@simcona.com
HTTP://www.simcona.com/
RATING: (7)

Simon Bridger Design
Designs equipment/systems
INFO: PIC info, links, project database
EMAIL: sbridger@world.std.com
HTTP://www.i-max.co.nz/sbridger/
RATING: (9)

Simple Technology
Memory and PC card manufacturer
INFO: Online memory configurer, drivers, etc.
EMAIL: info@simpletech.com
HTTP://www.simpletech.com/
RATING: (10)

Simtek Corp.
Manufactures nvSRAMs
INFO: Products, datasheets
EMAIL: See site for form
HTTP://www.csn.net/simtek/
RATING: (7)

Sipex Corp.
Analog IC manufacturer
INFO: Products, datasheets, cross ref
EMAIL: marketing@sipex.com
HTTP://www.sipex.com/
RATING: (8)

Sirius microSystems
PIC development products
INFO: Product listings, support & files
EMAIL: support@siriusmicro.com
HTTP://siriusmicro.com/
RATING: (8)

Sloan
Manufactures indicator lights and optoelectronic assems.
INFO: Company info, contact
EMAIL: sloancomp@aol.com
HTTP://www.d-b.com/sloan/
HTTP://eemonline.com.sloan/
RATING: (n/a)

SMT Plus
SMT training courses, etc.
INFO: Training info, links
EMAIL: jimb@smtplus.com
HTTP://www.smtplus.com/
RATING: (7)

SMTnet, Inc.
SMT, PCB and electronics manufacturing information supersite
INFO: EVERYTHING including manufacturers, orgs, forums, etc.
EMAIL: info@smtnet.com
HTTP://www.smtnet.com/
RATING: (10) **********

Society for Amateur Scientists
INFO: Member info, tech info galore!

S

EMAIL: info@sas.org
HTTP://www.thesphere.com/sas/
RATING: (10)

Softaid, Inc.
Embedded emulators
INFO: Just about every resource for
embedded processors
EMAIL: emulate@softaid.net
HTTP://www.softaid.com/
RATING: (10)

SoftSmiths Pty. Ltd. — Australia
Software for VLSI design
INFO: Product info, links
EMAIL: info@softsmiths.oz.au
HTTP://www.webventures.com.au/
ElectTech/softsmiths/
RATING: (7)

Sola Electric
HTTP://www.industry.net/
sola.electric/

Solid Electric, Inc. (SEI)
Manufactures connectors for PCB/
cable mount apps.
INFO: Catalog, links, service
EMAIL: techinfo@seinet.com
HTTP://www.seinet.com/
RATING: (9)

Solid State Technology
Online magazine, info site for
semiconductor engineers
INFO: Database, free subscription,
jobs
EMAIL: See site for forms
HTTP://www.solid-state.com/
RATING: (10)

Sony Corporation of America
Consumer electronics manufacturer,
music, movies, etc.
INFO: Connection to all of Sony's
divisions, news
EMAIL: sonyonline@sonyusa.com
HTTP://www.sony.com/
RATING: (Over-all: 10)

Sony Electronics, Inc.
Division of Sony Corp. of America
INFO: Semiconductor, consumer
business and professional products
EMAIL: contact@sel.sony.com
HTTP://www.sel.sony.com/sel/sel/
RATING: (Over-all 10)

Sony Semiconductor Company of America
Division of Sony, handling the
manufacture of semiconductors
INFO: Past, present, future product
info/datasheets, etc.
EMAIL:
ssaweb@ccmail.nhq.sony.com
HTTP://cons4.sel.sony.com/semi/
RATING: (10)

Sorenson Lighted Controls (SoLiCo)
Manufactures indicator lights/panel
LEDs
INFO: Products, reps, forms
EMAIL: See site for form
HTTP://solico.com/
RATING: (6)

Sound Marketing Industrial Electronics (SPI)
Canadian distributor
INFO: Product list
EMAIL: deasingwood@dowco.com
HTTP://www.christelle.com/smi/
RATING: (3)

Sound Radio Products
Radios and equipment distributor
INFO: Catalog with pics and prices
EMAIL: soundradio@aol.com
HTTP://www.eskimo.com/
%7Eantenna/
RATING: (8)

SoundSite, Inc.
Audio-related organization
INFO: Audio info, links, help
EMAIL:
webmaster@soundsite.com
HTTP://www.soundsite.com/
RATING: (10)

Southwest Electronic Energy Corp.
Value-added distributor
INFO: Products carried, requests
EMAIL: info@swe.com
HTTP://www.swe.com/
RATING: (4)

Space Electronics, Inc.
Extreme-Enviro ICs
INFO: Catalog, FAQ, news
EMAIL:
102005.1635@compuserve.com
HTTP://www.newspace.com/
industry/spaceelec/
RATING: (7)

Spacecraft Components Corp.
Connects solutions
INFO: Full product lists, links,
everything you wanted to know
about connectors but were afraid to
ask.
EMAIL: space@spacecraft.com
HTTP://www.spacecraft.com/
RATING: (9)

SPARC International, Inc.
Organization promotes SPARC
tech.
INFO: Membership info
EMAIL: webinfo@sparc.com
HTTP://www.sparc.com/
RATING: (7)

Spectra Test Equipment, Inc.
Buy/sell
INFO: Specials, inventory, links,
etc.
EMAIL:
paulhall@mail.spectratest.com
HTTP://spectratest.com/
RATING: (7)

Spectrum Control, Inc.
Manufactures EMC filter products
INFO: Product info, reps, news
EMAIL:
spectrum@spectrumcontrol.com
HTTP://www.spectrumcontrol.com/
RATING: (8)

Spire Corp.
EMAIL: spirecorp@channell.com

Sprague
Manufactures capacitors
INFO: Product info/datasheets
EMAIL: See site for FAX info
HTTP://vishay.com/vishay/sprague
RATING: (6)

Square D
HTTP://www.industry.net/squared

SR Huntting, Inc.
RF data communications software
INFO: KaWin info/software,
registration

S

EMAIL: stan@mutadv.com
HTTP://www.mutadv.com/kawin/
RATING: (7)

SSAC — Solid State Advanced Controls

Manufactures electronics controls
INFO: Product info/pics, reps, news
EMAIL: info@ssac.com
HTTP://www.ssac.com/
RATING: (7)

Stac Electronics

Network software/hardware
INFO: Service, products, events, etc.
EMAIL: See site for form
HTTP://www.stac.com/
RATING: (9)

Stacom Corporation

Wholesale independent electronics distributor
INFO: Specialty milspec parts
EMAIL: stacom@eazy.net
HTTP://eazy.net/stacom/
RATING: (6)

Stamp Applications

SEE: *Scott Edwards Electronics*

Stamp Applications

Bob Blick's homepage
INFO: Projects and links for the Basic Stamp — also see his other electronics projects site.
EMAIL: bob@ert.com
HTTP://www.bobblick.com/bob/stamp/index.html
RATING: (8)

Standard Microsystems Corp. (SMC)

Components Products Division
Semiconductor manufacturer
INFO: Component info/specs/support, drivers, library, you name it!
EMAIL: chipinfo@smc.com
HTTP://www.smc.com/
FTP://ftp.smc.com/
BBS: 516-273-4936
RATING: (10)

Stanford Microdevices

Manufactures GaAs FETs/MMICs
INFO: Product info/datasheets, distributors
EMAIL: dan_jensen@stanfordmicro.com
HTTP://www.stanfordmicro.com/
RATING: (9)

Stargate Connections, Inc.

ISP/Host for technology-related companies
INFO: Directory and links to companies, news, conferences, etc.
EMAIL: info@starcon.com
HTTP://www.starcon.com/
RATING: (10)

Stark Wholesale Electronics

Main TV antenna supplies
INFO: Product info, links
EMAIL: starkel@ma.ultranet.com
HTTP://www.ultranet.com/~starkel/
RATING: (7)

Static Power Conversion Services, Inc.

Source for power quality excellence
INFO: Company/service info, industry news
EMAIL: spcs@ix.netcom.com

HTTP://www.peinet.com/
staticpower
RATING: (6)

Steinhoff Automations- & Feldbus-Systeme
PC-based open control systems
INFO: Product info, drivers, PC/
104 info
EMAIL: info@steinhoff.de
HTTP://www.steinhoff.de
RATING: (8)

Sterling Electronics Corp.
Value-added distributor
INFO: Linecard, magazine, vendor
services
EMAIL: See site
HTTP://www.sterlink.com/
RATING: (8)

Sumitomo Electric USA, Inc.
Wire, navigation systems, etc.
INFO: Links to SEI divisions,
products, library, etc.
EMAIL: info@sumitomo.com
HTTP://seusa.sumitomo.com/
RATING: (7)

Summit Device Technology
IC design services
INFO: Service info
EMAIL: sdtech@verinet.com
HTTP://www.verinet.com/~sdtech/
RATING: (1)

Summit Distributors, Inc.
Electronics component distributors
INFO: Links, linecard, quotes
EMAIL: summitdist@aol.com
HTTP://members.aol.com/
summitdist/
RATING: (8)

Sun Light Electronics, Inc.
IC and component distributor
INFO: Linecard, contact
EMAIL: eicsun@aol.com
HTTP://www.sunlightusa.com/
RATING: (6)

Sun Microsystems, Inc.
RISC microprocessor manufacturer
INFO: Product info/support/sales/
tech, you name it!
EMAIL: See site for locations
HTTP://www.sun.com/
RATING: (10) ***

SUN's Design Automation Cafe
INFO: EDA info, jobs, news, talk,
university, tech
EMAIL: wwwadmin@dacafe.com
HTTP://www.dacafe.com/
RATING: (10)

Sunshine Wire & Connector, Inc.
Distributor
INFO: Line info, specials
EMAIL: sunshi15@ix.netcom.com
HTTP://www.sunshinewire.com/
RATING: (8)

Suntech Industries, Inc.
Distributes ICs, connects, memory
INFO: Linecard, reps
EMAIL:
76065.277@compuserve.com
HTTP://ourworld.compuserve.com/
homepages/sii_suntech/
RATING: (3)

S - T

Superior Electric
HTTP://www.industry.net/
superior.electric

SuperPower, Inc.
Power systems for industry
INFO: Links, product specs, profile
EMAIL: n/a
HTTP://www.superpower.com/
RATING: (7)

Supertex, Inc.
EMAIL: prodinfo@supx.com

Surplus Shack
Buys/sells optical and electronics
surplus
INFO: Specials
EMAIL: surplushak@aol.com
HTTP://www.SurplusShack.com
RATING: (5)

Surplus Traders
Caters to manufac./exporters
600 page on-line buy/sell catalog
EMAIL: marv@73.com,
ted@73.com
HTTP://www.73.com/a/
RATING: (8)

Surtek Industries, Inc.
Assembles/tests PCBs
INFO: Service info, contact
EMAIL: surtek-info@starcon.com
HTTP://www.starcon.com/surtek/
RATING: (6)

Switchcraft
Raytheon Subsidiary
INFO: Product/company info, pics
EMAIL: See site form

HTTP://www.raytheon.com/re/swc/
RATING: (7)

Switches Plus
Factory-direct switch supplier
INFO: Factory-direct online
catalog including products/pricing
EMAIL: info@switchesplus.com
HTTP://www.switchesplus.com/
RATING: (8)

Sylva Control Systems
Basic language programmable
controllers
INFO: Catalog, prices, download
EMAIL: sylva@mail.baynet.net
HTTP://
www.sylvacontrol.baynet.net
RATING: (7)

Symbios Logic, Inc.
Storage, ASICs, boards, etc.
INFO: Tools, libraries, product info
galore!
EMAIL: webmaster@symbios.com
HTTP://www.symbios.com/
FTP://ftp.symbios.com/pub/
RATING: (10)

Synergetics
SEE: *Guru's Lair*

Synergy Semiconductor Corp.
Manufactures high-speed/mixed-
signal ICs
INFO: Product info/datasheets,
sales, contact, etc.
EMAIL: info@synergysemi.com
HTTP://www.synergysemi.com/
RATING: (9)

T.A.K. DesignS
PIC microcontroller FAQ and links
INFO: Awesome reference for PICs
EMAIL: takdesign@digiserve.com

HTTP://digiserve.com/takdesign/
RATING: (10)

Taiwan Semiconductor
Manufacturer of VLSIs & ASICs
INFO: Company info, contact
EMAIL: See listing for nearest rep
HTTP://www.tsmc.com.tw/
RATING: (7)

Tandy Corp.
Radio Shack, Computer City and
Incredible Universe
INFO: New links, store locators,
online shopping, everything
imaginable between the 3 sites
EMAIL: See corporate contacts for
listing
HTTP://www.tandy.com/
RATING: (10)

Target Electronics, Inc.
HTTP://www.industry.net/
target.tron

Tauber Electronics, Inc.
Distributes batteries
INFO: Linecard, contact
EMAIL: batteri@electriciti.com
HTTP://www.tauber.com/~batteri/
RATING: (4)

TDK Corp.
Recording media, components,
factory equipment, PCMIA
INFO: Extensive product info,
news, support
EMAIL: See site for listings
HTTP://www.tdk.com/
RATING: (10)

Team America
Surplus liquidators and closeouts
INFO: Specials, inventory, etc,
EMAIL: teamamer@ix.netcom.com

HTTP://www.vir.com/jam/
team.html
RATING: (7)

The Tech Store
Distributes technician's tools
INFO: Line info
EMAIL: tecstore2@aol.com
HTTP://www.webcreations.com/
techstore/
RATING: (6)

Techlock Distributing
Microwave & RF components
INFO: Linecard with pics/prices,
contact
EMAIL: kenny@erinet.com
HTTP://www.erinet.com/kenny/
RATING: (8)

Techni-Tool, Inc.
Tool supplier
INFO: Linecard with great info/
pics, contact
EMAIL: techtool@interserv.com
HTTP://www.techni-tool.com/
RATING: (9)

Technik, Inc.
Press-N-Peel
Printed circuits from laser print
INFO: Order samples, instructions,
links
EMAIL: techniks@chelsea.ios.com
HTTP://chelsea.ios.com/~techniks
RATING: (8)

Technimation, Inc.
Electronic design/prototyping
INFO: Under construction at
printing
EMAIL: cytodd@technimation.com

T

HTTP://www.technimation.com/
RATING: (n/a)

Technology Newsstand — includes:
Real-time Engineering Magazine
VMEbus Systems Magazine
VXIJournal
INFO: Great online mags for professionals
EMAIL: magpub@aol.com
HTTP://www.primenet.com/~magpub/
RATING: (10)

Tecknit
Manufacturer of EMI/RFI shielding products
INFO: Reps, product info
EMAIL: tecknit@tecknit.com
HTTP://tecknit.com/
RATING: (7)

Teka Interconnection Systems
Solder/Flux bearing lead technology - SMT
INFO: Product info, description
EMAIL: szgh88a@prodigy.com
HTTP://w3.bbsnet.com/teka/
RATING: (7)

Tekman
SEE: *Audiophiles*

Teknor Industrial Computers, Inc.
Industrial single-board computers
INFO: Product info, tech support, contacts
EMAIL: sales, support@teknor.com
HTTP://www.teknor.com/
RATING: (9)

Tektronix Instruments
Measurement, printers, video/networking manufacturer
INFO: Products info/apps, jobs, news, etc. A1 graphics!
EMAIL: See site for lisiting
HTTP://www.tek.com/
RATING: (10)

Tellurex Corp.
Thermoelectric cooler/heating device manufacturer
INFO: Products, technology info on no-moving-part cooling
EMAIL: tellurex@tellurex.com
HTTP://www.tellurex.com
RATING: (9)

Telogy, Inc.
Sells test equipment
INFO: Buy/sell test equipment help site
EMAIL: See form on site
HTTP://www.tecentral.com/
RATING: (8)

Telsat Communications Ltd.
Satellite TVRO importer/distributor, New Zealand
INFO: TVRO products/FAQ/links
EMAIL: telsat@intec.gen.nz
HTTP://www.nethomes.com/telsat/
RATING: (9)

Teltone Corp.
Telecommunications manufacturer
INFO: Product info/apps, software, news
EMAIL: info@teltone.com
HTTP://www.teltone.com/
RATING: (9)

Telulex, Inc.
Synthesized signal generator/analyzer manufacturer

INFO: Datasheets, software upgrades, apps
EMAIL: sales@telulex.com
HTTP://www.telulex.com/
RATING: (8)

Temic Semiconductors

Semiconductor manufacturer, Germany
INFO: Products info/specs/datasheets, etc.
EMAIL: webmaster@temic.de
HTTP://www.temic.de/
RATING: (9)

Temp-Pro, Inc.

Temperature-sensing products manufacturer
INFO: Profile, products, custom info
EMAIL: sales@temp-pro.com
HTTP://www.temp-pro.com/
RATING: (7)

Test Systems

INFO: Product info, specs
EMAIL: www@teradyne.com
HTTP://www.teradyne.com/
RATING: (7)

Tessco Technologies, Inc.

Supplies wireless service products
INFO: Customer/tech tips, literature
EMAIL: webmaster@tessco.com
HTTP://www.tessco.com/
RATING: (9)

Test Engineering Home Page (Electronics)

Page by Russell Rapport
INFO: Links to about every test engineer resource out there, plus more!

EMAIL: rrapport@eden.com
HTTP://www.eden.com/~rrapport/testeng.htm
RATING: (9)

TestEquity, Inc.

New/refurbished test equipment
INFO: Linecard with specs/pics, contact
EMAIL: sales@testequity.com
HTTP://www.testequity.com/
RATING: (9)

Testforce Systems, Inc.

Distributor of test equipment in Canada
INFO: Linecard, locations
EMAIL: sales@testforce.com
HTTP://www.testforce.com/
RATING: (7)

Texas Instruments

Semiconductors, consumer products, etc.
INFO: Full of product data/apps/specs/etc. For TI products
EMAIL: See site for listing/form
HTTP://www.ti.com/
FTP://ftp.ti.com (192.94.94.3)
DSP FTP://140.111.1.10 directory/vendors/ti/tms320bbs
RATING: (10)

Thermalloy, Inc.

Heat sink manufacturer
INFO: Products, reps, literature, software
EMAIL: wnjf66c@prodigy.com
HTTP://www.thermalloy.com/
RATING: (7)

T

Thesys Microelectronics
German IC manufacturer
INFO: Jobs, products, contact
EMAIL: info@thesys.de
HTTP://www.thesys.de/
RATING: (7)

Thomas Register Homepage
The online version of *The Thomas Register*
INFO: A directory of manufacturers. Good for address/phone/fax searches
EMAIL: See site for form
HTTP://www.thomasregister.com:8000/
RATING: (9)

Tiare Publications
Books for radio communications hobbyists
INFO: Catalog, tips for amateur radio
EMAIL: info@tiare.com
HTTP://www.tiare.com/
RATING: (8)

Time Electronics, Canada
Avnet subsidiary distributing electronics components
INFO: Linecard, newsletter, jobs
EMAIL: See Avnet
HTTP://www.time.avnet.com/
RATING: (7)

Timeline, Inc.
Buys/sells LCDs and other electronics/computer items
INFO: Specials, hackers corner
EMAIL: n/a
HTTP://www.netrix.net/timeline/
RATING: (8)

Times Fiber Communications, Inc.
Coax and specially cable manufacturer
INFO: Company/product info/contact
EMAIL: 73424.1276@compuserve.com
HTTP://www.timesfiber.com/
RATING: (6)

Timewave Technology, Inc.
DSP products for amateur radio/industry
INFO: Company/product info/docs, links
EMAIL: dsp@timewave.com
HTTP://www.timewave.com
RATING: (9)

TLC Electronics, Inc.
Distibutor/Rep
INFO: Linecard
EMAIL: tlc@electronet.com
HTTP://www.electronet.com/tlcelect.htm
RATING: (3)

TLJ Consulting
SEE: *Filip Gieszczykiewicz's Sites*

Tokin America, Inc.
EMIs, filters, caps, etc., manufacturer
INFO: Product info/support, reps
EMAIL: See site listing for department
HTTP://www.tokin.com/
RATING: (8)

Toko America
Manufacturer of semiconductors/components
INFO: Product info/specs, news
EMAIL: info@tokoam.com

HTTP://www.tokoam.com/
RATING: (8)

Tomi Engdahl
Homepage with electronics links/
projects
INFO: Links, projects, everything a
hobbyist/professional needs
EMAIL: tomi.engdahl@iki.fi
HTTP://www.hut.fi/~then/
RATING: (10) ***

Topdown Design Solutions, Inc.
VHDL tools/models/training
INFO: Tools, training, support,
FAQs, etc.
EMAIL: webmaster@topdown.com
HTTP://www.topdown.com/
RATING: (9)

Toshiba America Electronics Components, Inc.
Consumer electronics,
semiconductors, etc.
INFO: Product/company info/specs,
you name it!
EMAIL: See site for listing
HTTP://www.toshiba.com/
RATING: (9)

Tracan Electronics Corp.
Canadian distributor/rep
INFO: Online catalog, support,
links
EMAIL: sales@tracan.com
HTTP://www.tracan.com/
RATING: (9)

Transcat
Test/calibration instruments
distributor
INFO: Linecard, newsgroups, tech
notes, online orders

EMAIL: webmaster@transcat.com
HTTP://www.transcat.com/
RATING: (10)

Transistor Devices, Inc.
Power supply manufacturer
INFO: Product info/pics
EMAIL: info@mailer.transdev.com
HTTP://www.transdev.com/
RATING: (7)

Transtronics, Inc.
Industrial electronics
INFO: Plenty of specs and product
info
EMAIL: info@xtronics.com
HTTP://www.xtronics.com/
& Electronic Hobby Kits
INFO: Electronic art, test
equipment, solder-less
EMAIL: kits@xtronics.com
HTTP://www.xtronics.com/kits.htm
RATING OVER-ALL: (8)

TranSwitch Corp.
VLSI semiconductors for
communications, etc.
INFO: Product datasheets/apps,
sales, support
EMAIL: mktg@txc.com
HTTP://www.transwitch.com/
RATING: (9)

Trendsetter Electronics
Distributes electronics components
I NFO: Catalog, links, contact
EMAIL: Form only
HTTP://www.trendsetter.com/
RATING: (5)

T

Trilithic, Inc.
RF equipment/component
manufacturer
INFO: Product info, orders info
EMAIL: sales@trilithic
HTTP://www.trilithic.com/
RATING: (8)

Triplett Corp.
Manufactures test equipment
INFO: Product info/pricing, news,
etc.
EMAIL: See site for form
HTTP://www.triplett.com
RATING: (9)

TriQuint Semiconductor, Inc.
Communications semiconductors
INFO: One page
EMAIL: sales@tqs.com
HTTP://www.triquint.com/
RATING: (2)

TRISYS, Inc.
Products and services for PICs
INFO: Software info, free help files
EMAIL: trisys@primenet.com
HTTP://www1.primenet.com/
~trisys/
RATING: (7)

TRS Consultants
Radio hobbyists, writer, Web page
design
INFO: Contributing editor, ACP
Symes "Radio and
Communications," shortwave
listening, amateur radio,
international broadcasting, very
comprehensive links page, news
and more
EMAIL: trs@trsc.com

HTTP://www.trsc.com/
RATING: (10)

TRW, Inc.
Automotive, space, military
technology
INFO: Product/corp info/apps, jobs,
you name it!
EMAIL: webmaster@trw.com
HTTP://www.trw.com/
RATING: (9)

TTI, Inc.
Distributor of passives &
interconnects
INFO: Locations, linecard
EMAIL: n/a
HTTP://www.ttiinc.com/
RATING: (7)

Tucker Electronics
Distributor of new & reconditioned
test & measurement equipment &
tools
INFO: Buy/sell/trade info, links,
news, you name it!
EMAIL: sales@tucker.com
HTTP://www.tucker.com/
RATING: (10)

Tundra Semiconductor Corp.
Bus-bridging and encryption
components
INFO: Product info/support/apps,
etc.
EMAIL: inquire@tundra.com
HTTP://www.tundra.com/tundra
RATING: (9)

TurboSim
Island Logics' seemless electronics
simulation
INFO: Download demo, offers,
product info

EMAIL: ljsn87a@prodigy.com
HTTP://pages.prodigy.com/L/J/A/
LJSN87A/
RATING: (7)

Tusonix, Inc.
EMI/RFI ceramic electronic
components
INFO: Product info/specs/cross ref.
EMAIL: sales@tusonix.com
HTTP://tusonix.com/
RATING: (9)

Twin Industries
PCB prototyping/production/
products
INFO: Product info/specs/pricing
EMAIL: twin@ix.netcom.com
HTTP://www.nilva.com/hunter/
twin1.htm
RATING: (7)

U.S. Army Missile Command
Real-Time Executive for
Multiprocessor Systems (RTEMS)
home page. Real-time O/S for
military purposes
INFO: Documentation, support,
links, download RTEMS
EMAIL: rtems@redstone.army.mil
HTTP://www.rtems.army.mil/
rtems.html
RATNG: (9)

U.S. Patent & Trademark Office
INFO: Search patent database,
download forms, etc.
EMAIL: N/A
HTTP://www.uspto.gov/
RATING: (10)

U.S. Robotics
Modems manufacturer
INFO: Products, support

EMAIL: See site for listing
HTTP://www.usr.com/
RATING: (9)

U.S. Sensor Corp.
Sensor manufacturer
INFO: Product info/specs/tech/pics
EMAIL: ussensor@ix.netcom.com
HTTP://www.ussensor.com/
RATING: (9)

UL — Underwriters Laboratories, Inc.
Product certification
INFO: Primer that takes you
through getting a UL mark
EMAIL: corpcomm@ul.com
HTTP://www.ul.com/
RATING: (10)

UN-L Engineering Electronics Shop
Internet resources and other EE
subjects
INFO: EE labs, software, project
stuff, etc. Wonderful page if you are
searching for projects!
EMAIL: mpaul@unlinfo.unl.edu
HTTP://engr-www.unl.edu/ee/
eeshop/
RATING: (10) ***

United Technology Microelectronics Center
Military and aerospace
semiconductors
INFO: Products, sales, news,
contact
EMAIL: farris@utmc.utc.com
HTTP://www.utmc.com/
RATING: (7)

U - V

University of Alberta Electrical and Computer Engineering
University site
INFO: EE-related info. See also Circuit Cookbook
EMAIL: web-master@ee.ualberta.ca
HTTP://www.ee.ualberta.ca/
RATING: (Over-all: 10)

Universal Systems Components
Stocking distributor, locator service
INFO: Database, specials, contact
EMAIL: usc@univsys.com
HTTP://www.univsys.com/
RATING: (5)

Unusual Diodes
Diode FAQ
INFO: Everything about diodes
EMAIL: mjc@avtechpulse.com
HTTP://www.avtechpulse.com/faq.html
RATING: (10)

URS Electronics, Inc.
Broadline distributor
INFO: Linecard, contact, links
EMAIL: sales@ursele.com
HTTP://www.ursele.com/
RATING: (7)

V3 Corp.
Embedded chipset company
INFO: Product briefs/apps/etc., design, contact, links
EMAIL: v3info@vcubed.com
HTTP://www.vcubed.com/
RATING: (10)

Valley Industrial Parts
Distributor of opto electronics
INFO: Specials, inventory, contacts
EMAIL: See site for sales reps
HTTP://www.av.qnet.com/~vip/
RATING: (6)

Valor Electronics, Inc.
(A GTI Company)
High-speed components manufacturer
INFO: Products, reps, literature
EMAIL: mktg@valorinc.com
HTTP://www.valorinc.com/
RATING: (9)

Valpey-Fisher
Crystal manufacturer
INFO: EEM online site/product data
EMAIL: n/a
HTTP://www.eemonline.com/valpey
RATING: (4)

VAutomation, Inc.
Synthesizable HDL cores
INFO: FAQs, links, product info
EMAIL: sales@vautomation.com
HTTP://www.vautomation.com
RATING: (8)

Vector Electronic Company
Developers of system packaging, prototyping
INFO: Reps, product info, catalog order
EMAIL: inquire@vectorelect.com
HTTP://www.vectorelect.com/
RATING: (7)

Vectron Laboratories
Crystal oscillator manufacturer
INFO: Product info, guide/cat. request

V

EMAIL: vectron@vectronlabs.com
HTTP://www.vectronlabs.com/
RATING: (7)

Vectron Technologies, Inc.
Manufactures ceramic surface
mount devices
INFO: Selection guide
EMAIL: gordonj
60125@mcimail.com
HTTP://www.vectron-vti.com/
RATING: (4)

VeriBest, Inc.
SEE ALSO: *Intergraph*
Broad line supplier of WinNT EDA
solutions for PCB, ASIC, FPGA
and PLD design
INFO: Products, customer support
and EDA industry trends
EMAIL: info@veribest.com
HTTP://www.veribest.com/
RATING: (9)

Vero Electronics
Supplier of racks, cabinets, etc.
INFO: White papers, products,
contact
EMAIL: vero@vero-usa.com
HTTP://www.vero-usa.com/
RATING: (7)

The Vertox Company
Panel meters/portable instrument
enclosures
INFO: Products/company info
EMAIL: vertox@1stweb.com
HTTP://www.1stweb.com/vertox/
RATING: (4)

VIA Technologies, Inc.
Semiconductor core logic
manufacturer
INFO: News, product info/
datasheets, support

HTTP://www.via.com.tw/
FTP://ftp.via.com.tw/
RATING: (9)

Vibro World
Custom electronics/tube devices
INFO: Technicians/custom shop,
links, you name it!
EMAIL: vibroman@teleport.com
HTTP://www.vibroworld.com/
~vibroman
RATING: (10)

Vicor Corp. (Express)
Power solutions manufacturer
INFO: Links, product, drawings
EMAIL: See online form
HTTP://www.vicr.com/
RATING: (9)

Vidicomp, Inc.
Dealer of professional/broadcast
video systems
INFO: Linecard, links, lists
products
EMAIL:
webmaster@vidicomp.com
HTTP://www.vidicomp.com/
RATING: (8)

Viewlogic Systems, Inc.
EDA software
INFO: Product info/support/
training, seminars
EMAIL:
viewdirect@viewlogic.com
HTTP://www.viewlogic.com/
RATING: (9)

Virtual Computer Corporation (VCC)
Reconfigurable computers

V - W

INFO: FAQs, links, papers, product info. Great information site
EMAIL: info@vcc.com
HTTP://204.58.152.114
RATING: (9)

The Virtual Hamfest

Web Publishing Co. — Classifieds
INFO: Ham events/classifieds/stolen
EMAIL: stevek@vhamfest.com
HTTP://www.webcom.com/webpub/
RATING: (9)

Virtual i-O

Manufacturer of head-mounted displays
INFO: Product info, dealers, 'Ask iO', you name it!
EMAIL: infor@vio.com
HTTP://www.vio.com/
RATING: (10)

The Vision and Imaging Technology Resource

Vision 1's information page for machine vision
INFO: Machine vision FAQs, links, software
EMAIL: See site for contact forms
HTTP://www.vision1.com/
RATING: (10)

Visionics

EDA software
INFO: Product info/reps/distributors, demos
EMAIL: visionics@bahnhof.se
HTTP://www.bahnhof.se/~visionics/
FTP: see site for link
RATING: (10)

VisionTek

Memory products manufacturer
INFO: News, online services, products, drivers, you name it!
EMAIL: webmaster@visiontek.com
HTTP://www.visiontek.com/
RATING: (10)

VITA - The VMEbus International Trade Association

INFO: Virtual trade shows, journals, links and other association features
EMAIL: marcom@vita.com
HTTP://www.vita.com/
RATING: (10)

Vitel Electronics

Tech solutions for OEMs in Canada
INFO: Linecard, links, products
EMAIL: See site for listing
HTTP://www.vitelelectronics.com/
RATING: (9)

VLSI Technology

ASICs/custom IC manufacturer
INFO: Server down during review
EMAIL: n/a
HTTP://www.vlsi.com/
RATING: (n/a)

Voltronics Corp.

Manufactures trimmer capacitors
INFO: Request info, "pick your trimmer"
EMAIL: voltron@styx.ios.com
HTTP://www.voltronicscorp.com/
RATING: (7)

W.W. Grainger, Inc.

Distributor of industrial and commercial equipment and supplies
INFO: Online catalog/ordering, services, store locator

EMAIL: feedback@grainger.com
HTTP://www.grainger.com/
RATING: (9)

Wago
Terminal blocks manufacturer
HTTP://www.industry.net/wago

Wakefield Engineering
Manufactures heatsinks, etc.
INFO: Products/services
EMAIL: n/a
HTTP://www.wakefield.com/
RATING: (5)

Wavetek
Test/measurement equipment
manufacturer
INFO: Product info/datasheets/pics.
Very well done site!
EMAIL:
webmistress@wavetek.com
HTTP://www.wavetek.com/
RATING: (10)

WE-MAN's Homepage
Electronics-related homepage
INFO: Tons of electronics info to
keep your browser busy!
EMAIL:
s.wieman@student.utwente.nl
HTTP://
cal003109.student.utwente.nl/
stefan/
RATING: (10)

Webb Distribution, Inc. (WDI)
Industry distributor of various
components
INFO: Linecard, links
EMAIL: sales@webbdistinc.com
HTTP://www.webbdistinc.com/
RATING: (8)

W

Webb Laboratories
RF & microwave CAE software
INFO: Software descriptions,
pricing, links
EMAIL: info@webblabs.com
HTTP://www.webblabs.com/
RATING: (8)

webTrader
Tiffany Associates sponsored
semiconductor manufacturing
equipment classifieds
INFO: Classifieds with pics
EMAIL: rtiffany@ix.netcom.com
HTTP://webtrader.com/webtrader/
RATING: (8)

Weidmuller, Ltd.
Designer of industrial hardware
INFO: Product info/specs, contacts
EMAIL: info1@weidmuller.ca
HTTP://www.weidmuller.ca/
RATING: (7)

Wenzel Associates
Manufacturer of crystal oscillators
INFO: Catalog, newsletter, links,
circuit library
EMAIL: wenzel@wenzel.com
HTTP://www.wenzel.com
RATING: (10)

Wes-Garde Components Group
Distributor specializing in electro-
mechanical components
INFO: Linecard, locations
EMAIL: See site for form
HTTP://www.wesgarde.com/
RATING: (7)

W

Wescorp
Static control products
INFO: Profile, products, newsletter
EMAIL:
fruhar@wescorpstaticcontrol.com
HTTP://
www.wescorpstaticcontrol.com/
RATING: (8)

Western Digital Corp.
Harddrives/computer components
INFO: Product info/support/tips,
reps, EVERYTHING!
EMAIL: webmaster@wdc.com
HTTP://www.wdc.com/
FTP://ftp.wdc.com
RATING: (10)

Western Micro Technology, Inc.
National distributor
INFO: Company, locations, line
EMAIL:
sysadmin@westernmicro.com
HTTP://www.westernmicro.com
RATING: (7)

Westinghouse Electric Corp.
Westinghouse and their divisions
INFO: Links to their products/
divisions
EMAIL: See divisions for listing
HTTP://www.westinghouse.com
RATING: (8)

WG Communications, Inc. (Wandel & Goltermann)
Communications test equipment
manufacturer
INFO: Product info/pics/data/
software
EMAIL: See site for form

HTTP://www.wg.com/
RATING: (8)

Wickmann
Manufactures circuit protection
components
INFO: Product info/specs, fuse
facts, distributors
EMAIL:
service@wickmannusa.com
HTTP://www.wickmannusa.com/
RATING: (9)

Wieland
Manufactures terminal blocks
HTTP://www.industry.net/wieland

Wilson Greatbatch, Ltd. & Subsidiaries
Batteries & product engineering
INFO: Product and subsidiary info
EMAIL: See site for listing
HTTP://www.greatbatch.com/
RATING: (7)

Winbond Electronics Corp.
Manufactures RAMs, uCs, ICs, etc.
INFO: Profile, products, sales
EMAIL: See site
HTTP://www.winbond.com.tw/
RATING: (9)

Winchester Electronics — Litton Systems, Inc.
Manufactures connectors
INFO: Product catalogs/datasheets,
news, jobs
EMAIL: techinfo@litton-wed.com
HTTP://www.litton-wed.com/
RATING: (8)

Wind River Systems
Embedded development systems
INFO: Datasheets, guide, white
papers

EMAIL: inquiries@wrs.com
HTTP://www.wrs.com/
RATING: (10)

Wisconsin Electronics Supply
Distributor
INFO: Linecard
EMAIL: wesinfo@wesfdl.com
HTTP://www.wesfdl.com/
RATING: (5)

Woodhead Company (Daniel)
SEE: *Daniel Woodhead Company*

World Wide Wire
Distributes wire, etc.
INFO: linecard, few links
EMAIL: wwwire@net5.com
HTTP://www.worldwidewire.com/
RATING: (3)

Worldwide Digital Connections, Inc.
Electronics design firm
INFO: Company info, contact
EMAIL: wwdc@onramp.net
HTTP://rampages.onramp.net/
~wwdc/
RATING: (4)

Worldwide Equipment Brokers
Online database of used
manufacturing equipment
INFO: Inventory database, list
EMAIL: info@worldeb.com
HTTP://www.worldeb.com/
RATING: (8)

The WWW Virtual Library, EE
HTTP://epims1.gsfc.nasa.gov/
engineering/ee.html
The author attempted 10
connections to this resource and
was unsuccessful.

W - X

Wyle Electronics — Wyle Ginsbury
Distributor of semiconductor,
computer products
INFO: Products linecard with links,
sales, support, take-a-break, you
name it! Great distribution site.
EMAIL: corpcom@wyle.com.
HTTP://www.wyle.com/
RATING: (10)

Xact, Inc.
Contract engineering specializing
in networking
INFO: Embedded info, product
info, fun
EMAIL: warren@xactinc.com
HTTP://www.xactinc.com
RATING: (8)

XECO — Now Integrated Process Systems, Inc.
Horizontal/vertical processing
equipment for electronics industries
INFO: Product/company info
EMAIL: mbrask@tcd.net
HTTP://www.tcd.net/~jstaudte/
RATING: (7)

XECOM, Inc.
Module modems for various apps
INFO: Specs, markets, sales
EMAIL: info@xecom.com
HTTP://www.rahul.net/xecom/
RATING: (8)

Xicor, Inc.
IC semiconductor manufacturer
INFO: Datasheets, apps, sales,
bulletins, you name it!
EMAIL: info@smtgate.xicor.com

X - Z

HTTP://www.xicor.com/
RATING: (10)

Xilinx, Inc.
Manufactures PLDs, FPGAs, etc.
INFO: Company, products, support,
programs
EMAIL: webmaster@xilinx.com
HTTP://www.xilinx.com/
FTP://ftp.xilinx.com
RATING: (10)

Yahoo — EE
HTTP://www.yahoo.com/science/
engineering/elecrical_engineering/

Yale Electronics
Distributes components/tools for
broadcast
INFO: Company info
EMAIL: sales@proyale.com
HTTP://www.proyale.com/
RATING: (4)

Yamaichi Electronics USA, Inc.
Manufactures IC sockets
INFO: Products, contact, RFQs
EMAIL: alanj@earthlink.net
HTTP://www.yeu.com/
RATING: (8)

Yuasa-GBC, Inc.
Battery/charger manufacturer
INFO: Product specs/pics/reps
EMAIL: See site for locations
HTTP://www.yuasagbc.com/
RATING: (7)

Z-Communications, Inc.
Voltage-controlled oscillator
manufacturer

INFO: Product selector, tech,
catalog, FAQ
EMAIL: sales@zcomm.com
HTTP://www.zcomm.com/
RATING: (9)

ZEN Archive
Zahir's electronic newsletter
INFO: Happenings in the
electronics community
EMAIL: zahir@eecg.utoronto.ca
HTTP://www.eecg.toronto.edu/
~zahir/
RATING: (8)

Zero Corp.
Specialty enclosures for electronics
INFO: Product/company info
EMAIL: zero@zerocorp.com
HTTP://www.zerocorp.com/
RATING: (6)

Zetex Semiconductors
Semiconductor manufacturer
INFO: Product datasheets/models,
news
EMAIL: See site
HTTP://www.zetex.com/
RATING: (7)

Ziatech Corp.
Compact PCI and STD 32
computers, etc.
INFO: Product support/apps/
settings
EMAIL: info@ziatech.com
HTTP://www.ziatech.com
RATING: (10)

Ziff-Davis Publishing Company
ZD Net — Computer Magazine
publisher

INFO: Publications, software library, community center. An A-Z internet site!
EMAIL: See site for form
HTTP://home.zdnet.com/
RATING: (10) *****

Zilog, Inc.
Microprocessor/support manufacturer
INFO: Product info/apps/specs, contacts, tools
EMAIL: See site for locations
HTTP://www.zilog.com/
RATING: (9)

Zorin
uC products/kits + extensions for automation & MIDI
INFO: Technical information, prices, orders, related links
EMAIL: info@ZORINco.com
HTTP://ZORINco.com
RATING: (9)

Zycad Corp. & Divisions
IC design simulation software, ASICs, FPGAs, prototyping
INFO: Verification, Gatefield, services div. info
EMAIL: Onsite form
HTTP://www.zycad.com/
RATING: (8)

ZZZAP Power Corp.
Distributes various power products
INFO: Line info, prices, contact
EMAIL: info@zzzap.com
HTTP://www.zzzap.com/
RATING: (8)

THE FUTURE OF ELECTRONICS
ON THE INTERNET

Every device controlled by a microchip will one day be connected to the Internet. Cars, TV's, stereos, home appliances,
even multimeters, will be talking to each other using a common language, forming one massive global electronics network.

Example: Your automobile's engine diagnostic computer will
be connected to the Internet. The dealer can diagnose and
treat its ailments using his web-browser, even though he is
2000 miles away. Example: the EPA is proposing a clean air
bill and wants comprehensive data of what cars are the
highest polluters. A quick connect to the Net will produce
specs from every car's engine diagnostic computer.

Networking is the logical evolution for computers. One computer can store X amount of data and process at a given
speed. 100 million computers can store 100,000,000 times
that data and process at equal speed but in parallel. Each
computer on the Net would process one task and pass it on
to another that would store a variable, etc.

Picture a large number of computers; not able to communicate with one another, going about their simple tasks. Now
give them the ability to talk to one another, transfer information back and forth, process it, and send it on. This is how
human knowledge started and is perpetuated. The Internet is
a knowledge network.

As an electronics technician of the future, you will need prior knowledge of the function of this new omnipresent Network. Who will design new applications? Who will program the new chips? Who will service new equipment?

NEAR FUTURE

Is the Internet a fad or will it become a common item on our workbench? What does the future hold for the Web? What does it hold for electronics on the Internet?

Electronics will be one of the few fields to survive the great put-up-or-shut up phase the Web is likely to go through. New modems will give rocketlike bandwidth, 35 times faster, almost overnight. Soon the Network Computer (NC), also dubbed Web PC, will allow anyone to connect at low-cost. Chips are being developed that will interact with the Internet, and software publishers are releasing extraordinary user features. Where will it end?

The Internet is changing the way we look at electronics. Easy access to data promises projects galore, and the need for technical support. The possibilities will rejuvenate the thrill you had when you made your first circuit. By steering toward the Information Superhighway's on-ramps early, you are ensuring a strong base knowledge of the Internet-of-the-future — an electronics Internet.

The future is awaiting us.

APPENDIX A:
IEEE DIRECTORY

IEEE — Institute of Electrical and Electronics Engineers
EMAIL: See site for directory
HTTP://www.ieee.org/

Aerospace and Electronic Systems Society
EMAIL: bob.trebits@gtri.gatech.edu
HTTP://rsd1000.gatech.edu/public_html/bobt/
AESS_Home_Page.html

Circuits and Systems Society
EMAIL: cas@www.ee.gatech.edu
HTTP://www.ee.gatech.edu/orgs/cas/

Communications Society
EMAIL: See site
HTTP://www.ieee.org/comsoc/comsochome.html

Components, Packaging, and Manufacturing Technology Society
EMAIL: w.trybula@ieee.org
HTTP://naftalab.bus.utexas.edu/~ieeecpmt/

The World's Computer Society
EMAIL: webmaster@computer.org
HTTP://www.computer.org/

Consumer Electronics Society
EMAIL: See site
HTTP://www.ieee.org/ce/

Control Systems Society
EMAIL: o.gonzalez@ieee.org
HTTP://www.odu.edu/~ieeecss/

Dielectrics and Electrical Insulation Society
EMAIL: n/a
HTTP://www.eng.rpi.edu/dept/epe/WWW/DEIS/

Education Society
EMAIL: b-oakley@uiuc.edu
HTTP://w3.scale.uiuc.edu/ieee_ed_soc/

Electromagnetic Compatibility Society
EMAIL: member.services@ieee.org
HTTP://www.emclab.umr.edu/ieee_emc/

Electron Devices Society
EMAIL: lgdm@ece.neu.edu
HTTP://www.ece.neu.edu/eds/EDShome.html

Engineering in Medicine and Biology Society
EMAIL: susan_blanchard@ncsu.edu
HTTP://www.bae.ncsu.edu/research/blanchard/www/embs/

Engineering Management Society
EMAIL: c.rubenstein@ieee.org
HTTP://sils.pratt.edu/796embog.html

Geoscience and Remote Sensing Society
EMAIL: tilton@chrpisis.gsfc.nasa.gov
HTTP://www.ieee.org/grs/index.html

Instrumentation and Measurement Society
EMAIL: o.carver@i ee.org
HTTP://www.ieee.org/im/www-3s.htm

Lasers and Electro-Optics Society
EMAIL: lah@msrc.wvu.edu
HTTP://msrc.wvu.edu/leos/

Magnetics Society
EMAIL: nyenhuis@ecn.purdue.edu
HTTP://yara.ecn.purdue.edu/~nyenhuis/ieeesmag.html

Microwave Theory and Techniques Society
EMAIL: soc.mtt@ieee.org
HTTP://www.ieee.org/mtt/mtt.html

Neural Networks Council
EMAIL: p.arabshahi@ieee.org
HTTP://www.ieee.org/nnc/

Oceanic Engineering Society
EMAIL: eric@cs.tamu.edu.
HTTP://auv-www.tamu.edu/oes/

Power Electronics Society
EMAIL: n/a
HTTP://www.ieee.org/society/pels

Power Engineering Society
EMAIL: See site
HTTP://www.ieee.org/power/power.html

Professional Communication Society
EMAIL: prslkg@arn.net
HTTP://www.ieee.org/pcs/pcsindex.html

Reliability Society
EMAIL: relpgm@eng.umd.edu
HTTP://www.enre.umd.edu//reinfo.htm

Robotics and Automation Society
EMAIL: n/a
HTTP://www.acim.usl.edu/RAS/

Signal Processing Society
EMAIL: sp.info@ieee.org
HTTP://ww.ieee.org/sp/index.html

Social Implications of Technology Society
EMAIL: j.herkert@ieee.org.
HTTP://www4.ncsu.edu/unity/users/j/jherkert/index.html

Systems, Man and Cybernetics Society
EMAIL: tg@isye.gatech.edu
HTTP://www.isye.gatech.edu/ieee-smc/

Ultrasonics, Ferroelectrics and Frequency Control Society
EMAIL: haddadin@eecs.umich.edu
HTTP://bul.eecs.umich.edu/uffc/

IEEE PUBLICATIONS

IEEE Institute
EMAIL: See site
HTTP://www.institute.ieee.org/ti.html

IEEE Potentials
EMAIL: See site
HTTP://www.cs.umr.edu/potentials/

IEEE Spectrum
EMAIL: spectrum-webmaster@ieee.org
HTTP://www.spectrum.ieee.org/

IEEE Transactions/Journals/Letters Preview
HTTP://www.ieee.org/pub_preview/pub_prev.html

IEEE Transactions on Semiconductor Technology Modeling and Simulation
HTTP://engine.ieee.org/journal/tcad/

APPENDIX B:
UNIVERSITIES and
UNIVERSITY ORGANIZATIONS

California Institute of Technology
EMAIL: See individual sites for addresses
HTTP://www.caltech.edu/

Computer Research Organization
EMAIL: webmaster@cra.org
HTTP://cra.org/

MICAS Research Group
EMAIL: n/a
HTTP://www.esat.kuleuven.ac.be/micas/

MIT — Massachusetts Institute of Technology
EMAIL: web-request@mit.edu
HTTP://web.mit.edu/

MIT Microsystems Technology Laboratories
EMAIL: boning@mtl.mit.edu
HTTP://www-mtl.mit.edu/

The National Center for Supercomputing Applications
University of Illinois at Urbana-Champaign
EMAIL: webdev@ncsa.uiuc.edu
HTTP://www.ncsa.uiuc.edu/

Stanford Department of Electrical Engineering
EMAIL: See individual sites for addresses
HTTP://www-ee.Stanford.edu/

University of Alberta — EE Department
EMAIL: See site.
HTTP://www.ee.ualberta.ca/

University of Auburn Electrical Engineering Department
EMAIL: See site for directory
HTTP://www.eng.auburn.edu/department/ee/eehome.html

University of California, Berkeley — Electrical Engineering & Computer Sciences
EMAIL: www-eecs@eecs.berkeley.edu
HTTP://www.eecs.berkeley.edu/

University of Idaho Microelectronics Research Center
EMAIL: lharold@mrc.uidaho.edu
HTTP://www.mrc.uidaho.edu:80/

University of Illinois Department of Electrical and Computer Engineering (ECE)
EMAIL: www@ece.uiuc.edu
HTTP://www.ece.uiuc.edu/

University of Maryland — Department of Electrical Engineering
EMAIL: ee-web-editor@eng.umd.edu
HTTP://www.ee.umd.edu/

U of Michigan Electrical Engineering and Computer Science Department
EMAIL: web@eecs.umich.edu
HTTP://www.eecs.umich.edu/

University of Sheffield — Department of Electronic & Electrical Engineering
EMAIL: j.screaton@sheffield.ac.uk
HTTP://www.shef.ac.uk/~eee/

University of Toronto — Electrical and Computer
EMAIL: See site
HTTP://www.ece.utoronto.ca/

University of Wisconsin-Madison — The College of Engineering
EMAIL: webmaster@engr.wisc.edu
HTTP://www.engr.wisc.edu/

Washington University in St. Louis — Electrical Engineering Department
EMAIL: webmaster@ee.wustl.edu
HTTP://ee.wustl.edu:80/ee/

For a complete list of World-Wide Universities, go to *Yahoo!* and use this path...

Top:Education:Higher Education:Colleges and Universities:

GLOSSARY OF TERMS

@. IRC: An operator of that channel. Example: @Rips
EMAIL: "at" Example: user@here.com

~ Tilde. Top-right key below the *Esc* key. Used in URLs usually indicating a user site. Example: HTTP:// www.site.com/~user/

404 or error 404. Could not find URL. Also means someone is a few cards short of a full deck.

ALT. The *alt* hierarchy on Usenet. The tree of newsgroups created by users without a formal vote or approval procedure. There is a myth, not entirely implausible, that *alt* is acronymic for "anarchists, lunatics, and terrorists"; but, in fact, it is simply short for *alternative*.

Apps. Short for *applications*. A term used to suggest the possible uses of a product. Can also mean a document that lists applications of a product. Example: The apps for this chip are listed below.

ARCHIE. A resource that allows you to search the Net for computer files with a specific file name.

ASCII. American Standard Code for Information Interchange. The character code of alphanumeric, punctuation, and control characters. It is represented by a number between 0 and 127 (7-bit number). Extended ASCII boosts the range to 0-255 (8-bit number). Example: Decimal number 65 or binary #1000001 is the letter A.

ASCII Schematics (ASCII ART). Drawings or electronic schematics made with ASCII characters.

ASICs. Application Specific Integrated Circuits. A chip that allows the designers to customize applications.

ATE. Automatic (automated) Test Equipment.

ATM. Asynchronous Transfer Mode. A cell-relay based network.

BITNET. A global network for academic institutes and re-search organizations. It is separate from the Internet.

Boolean. The logic that search engines and computers are based upon.

Boolean Operators. Used between keywords to narrow searches: AND, OR, NOT, and NEAR.

Browser or **Web-Browser**. The program used to access the World Wide Web. Example: Netscape Navigator or Microsoft Internet Explorer.

DIY. Do It Yourself. Example: *Buy our DIY kit*.

Driver. A program that controls a device. Companies provide these on their web site's tech support section.

DSO. Digital Storage Oscilloscopes. An oscilloscope that can store its information digitally.

DSP. Digital Signal Processor (Processing). A microproces-sor that can manipulate, correct, and modify signals digitally.

EDA. Electronic Design Automation. Software used to aid circuit and PCB board design.

EE. Electrical Engineer or Electronics Engineer

EMAIL. Electronic Mail. Messages or files sent electronically.

EMC. Electro Magnetic Compatibility.

FAQs. Frequently Asked Questions. A query/reply type primer to help people with common information about subjects on the Internet.

Flaming. When someone lashes out rudely over a usenet posting or Email of another. *Flame* as a noun means, "An insulting message".

FPGA. Field-Programmable Gate Array.

FTP. File Transfer Protocol. An Internet resource used to upload (send) and download (receive) files.

Fuzzy Logic. A logic system that uses multi-valued degrees of true/false as opposed to only on/off states. Sometimes referred to as *Grade Logic*.

GOPHER. An older Internet resource that operates on text only. It is still used for many databases, however.

HTML. HyperText Markup Language. The language used to create Web pages and link sites to one another.

HTTP. HyperText Transfer Protocol. The protocol that lets the your computer communicate with the Web. (Example: connect, request, respond, close.)

IEEE. Institute of Electrical and Electronics Engineers.

Intranet. An internal corporate network, similar to the Internet but not connected to it.

Internet Phone. A low-tech way to communicate with voice through the Net.

IRC. Internet Relay Chat. The Internet resource that allows you to converse with others around the world. It is text-based but is now moving to voice and limited video.

ISP. Internet Service Provider. The company which supplies your dial-up connection to the Internet.

Java. A programming language that lets small executable programs and other items be placed within a web page.

Linecard. A sheet or web page a distributor puts out that lists products they carry and notes the manufacturer; sometimes includes valuable links to each manufacturer.

Listserv. A *listserv* is a program located on the Internet server that processes commands received from outside sources. This is usually used to process and send mass Email such as a mailing list.

Mailing List. Similar to Usenet but done through Email. You subscribe to the list, usually through a *listserv*; then whenever anyone posts to it, each subscriber receives the post via Email.

MCU. Microcontroller Unit. A microprocessor with onboard I/O ports.

MPEG. Moving Picture Experts Group. A type of digital video compression.

MPU. Microprocessor Unit.

Newbie. Someone who is new to a field, such as the Internet.

Newsgroups. A resource on the Internet (basically a discussion group or bulletin board) containing articles and threads (responses) about any topic imaginable.

OEM. Original Equipment Manufacturer.

PCI. Peripheral Component Interconnect (local bus). The standard bus structure used in modern PCs.

PDF. Portable Document Format. A document format with multiple document types integrated within. Most datasheets on the Internet are in PDF files and need a PDF reader to view. See *Internet Resources* for further information.

PIC. The family of RISC microcontroller chips made by Microchip Inc.

PLD. Programmable Logic Device.

PPP. Point to Point Protocol. The computer protocol that manages your connection to the Internet. Your dial-up account will most likely use this protocol to allow your computer to communicate to the Network.

Rep. A product or company representative.

RFQ. Request For Quote.

RISC. Reduced Instruction Set Computer. A type of computer architecture that uses less cycles to execute computing tasks.

sci. The newsgroups that deal with science. Example: sci.electronics.basics

SCSI. Small Computer Systems Interface.

Search Engine. Programs that search through documents according to user defined variables.

SLIP. Serial Line Interface Protocol. Another type of dial-up account similar to PPP.

Smileys. :) :-) :-(etc. Look at them sideways. A way to show expression in messages. Used primarily in IRC and Usenet. For dictionaries of smileys, see: HTTP:// www.yahoo.com/Arts/Computer_Generated/ASCII_Art/ Smilies/

Spam(ming). Advertising in places where it is inappropriate.

Surfing. Wandering around the WWW aimlessly. Going from one web site to another without looking for anything in particular.

TCP/IP. Transmission Control Protocol and Internetwork Protocol. The universal protocol computers use to communicate to one another on the Internet.

TELNET. An older Internet resource that permits remote access to computers on the Net without having to call up that computers specific phone lines. This archaic service is relegated to mostly government and public libraries which require its use.

TVRO. TeleVision Receive Only.

uC. Short for Microcontroller.

UK. United Kingdom.

UNIX. A computer operating system. Found on many servers throughout the Internet.

uP. Short for Microprocessor.

URL. Uniform Resource Locator. The "address" or location of an Internet site/file. Can be an http://, or ftp://, or gopher://, or other information site. The URL can contain 1) protocol 2) domain name 3) path 4) file name. Example: http://www.site.com/directory/filename.html

Usenet. See newsgroups.

VAR. Value-Added Retailer

VHDL. VHSIC Hardware Description Language.

VLSI. Very Large-Scale Integration.

VMEbus. Versa Modula Europa BUS.

Web. The Web. Short for World Wide Web.

Web Client. Your web-browser. Example: Netscape Navigator, Microsoft Internet Explorer.

Web Master. The person in charge of that company's web site. Also called Web Mistress.

Web Page. An HTML document with various links and objects placed in it.

Web Server. The computer that the HTML documents reside on.

Web Site. An area on the web server which contains multiple Web pages.

Wild Card. Symbol(s) that takes the place of an unknown number of characters. Example: elec* for electronics or electrical, etc.

WWW. World Wide Web. The global resource that consists of networked computers containing HTML documents, all available for you to access with your PC.

INDEX

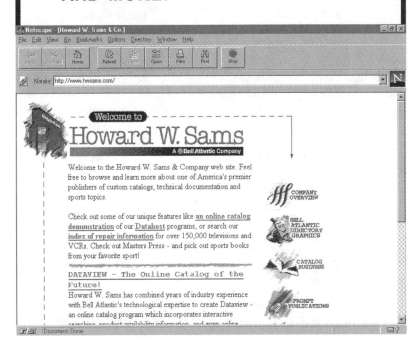

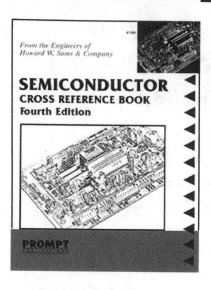

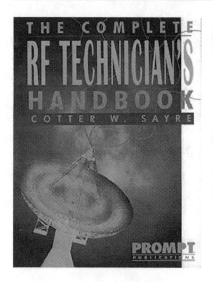
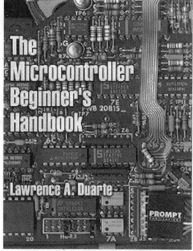

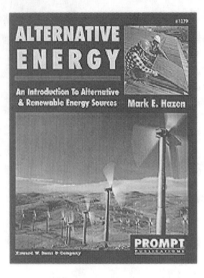